CLUMSY FLOODF

To my brother Martin

Clumsy Floodplains
Responsive Land Policy for Extreme Floods

THOMAS HARTMANN
Utrecht University, The Netherlands

Routledge
Taylor & Francis Group

LONDON AND NEW YORK

First published 2011 by Ashgate Publishing

Published 2016 by Routledge
2 Park Square, Milton Park, Abingdon, Oxfordshire OX14 4RN
711 Third Avenue, New York, NY 10017, USA

First issued in paperback 2016

Routledge is an imprint of the Taylor & Francis Group, an informa business

British Library Cataloguing in Publication Data
Hartmann, Thomas, 1979-
  Clumsy floodplains : responsive land policy for extreme
  floods.
  1. Floodplain management.
  I. Title
  363.3'4936-dc22

Library of Congress Cataloging-in-Publication Data
Hartmann, Thomas. 1979-
  Clumsy floodplains : responsive land policy for extreme floods / by Thomas Hartmann.
     p. cm.
  Includes bibliographical references and index.
  ISBN 978-1-4094-1845-0 (hbk)
  1. Floodplains. 2. Floodplain management. 3. Storm surges. 4. Flood damage prevention. I. Title.
  GB561.H37 2011
  363.34'936--dc22

                                                                        2010033267

ISBN 13: 978-1-138-25202-8 (pbk)
ISBN 13: 978-1-4094-1845-0 (hbk)

Contents

List of Figures

About the Author

Dr Thomas Hartmann is currently a post-doctral researcher at the Department of Human Geography and Spatial Planning, Utrecht University. In his research on environmental governance he combines engineering, socio-political research and anthropological studies to explore the relationship between land and water.

In 2009, he finished his PhD on 'Clumsy Floodplains – Responsive Land Policy for Extreme Floods' (Dr. rer.pol.) at the Department for Land Policy, Land Management and Municipal Geoinformation, TU Dortmund. Before, he studied Spatial Planning at the School of Spatial Planning, Dortmund. His diploma thesis was about 'Hochwasserschutz durch räumliche Planung – Handlungsempfehlungen für Abstimmungsprozesse zwischen den Akteuren in Sachsen-Anhalt' [Flood protection by spatial planning – recommendations for decision-making processes of stakeholders in Saxony-Anhalt] (Dipl.-Ing., 2005).

Thomas Hartmann appreciates comments and remarks on this book: t.hartmann@geo.uu.nl.

Preface
A Passion for Floods

The slogan 'Living near the River Elbe', on a signboard, promoted a new housing area in Dessau in August 2002. In the back, the brown water of the Elbe flood almost overflowed the sandbags. Volunteers like me put them on top of the levees in order to defend against the water. There were many voluntary helpers. I came with *Die Johanniter* (German St. John Ambulance) 600 km from Dortmund in the Ruhr area to protect these and other residential areas. We succeeded in Dessau. At the same time, my brother Martin was in Dresden with the *Technisches Hilfswerk*, THW (a German governmental organisation for disaster relief). He and his colleagues pumped floodwater out of the *Frauenkirche*. Like in other parts of Eastern Germany, in Dresden, the sandbags could not hinder the water inundating cities and landscapes. Dresden is upstream of Dessau. The inundation of the City of Dresden, the crevasses and unintended flooding of large areas between Dresden and Dessau helped us in Dessau to defend this new housing area 'living near the River Elbe'.

I began to wonder how society manages riparian floodplains. It seems that people have wanted to use floodplains for a long time. They settled there, built cities and industries, lived and enjoyed the riverside. However, sometimes it is dangerous in these areas. Society has found a way of responding to the elemental forces of the water in the floodplains. I found this way somewhat clumsy. For this reason, since 2002, I have focused my study of Spatial Planning on the flood issue. The relation between spatial planning and flood protection caught my interest in my diploma thesis (Hartmann 2005). I found no sufficient answer to explain how society could challenge the elemental water forces in the floodplains better. I became passionate about this challenge. This is the reason for this book.

Sorry, my dear parents, my brother Martin, my dear friends and colleagues for bothering you with talking all day long about floods, clumsiness, and Large Areas for Temporary Emergency Retention – which you all know as LATER. Thank you Professor Benjamin Davy and Professor Barrie Needham for supervising me (and encouraging me to write this book in English). Thanks also to Michael Thompson, for the inspiring discussions about Cultural Theory and Brigitte Hower, for all your support. Thank you, dear landowners, water managers, policymakers, and land use planners in the floodplains. This book would not have been possible without you. And sorry to all who believed I would stop talking about floods after my PhD, for I will disappoint you.

Thomas Hartmann
November 2010

Structure of the Book

How can society manage the use of floodplains along the rivers in the face of extreme floods? The relation between the social construction and environmental constraints of river floods is the overall interest of the main research question. Answering this question requires an interdisciplinary approach. Methods and theories from different disciplines and its combination help to answer the following three research questions: why is contemporary floodplain management clumsy? Which floodplain management can cope with extreme floods in the future? How can such floodplain management be implemented in clumsy floodplains?

Chapter 1, 'Clumsy Floodplains', answers the first of the three questions. Empirical observations and interviews with essential stakeholders in the floodplains illustrate the patterns of human activity in the floodplains. The semi-structured interviews have mainly been conducted in 2004 and 2008 along the River Elbe in Saxony-Anhalt and along the River Rhine in Cologne. Important statements of interviewees are written in the original language (*in italics*), whereas the meaning is translated into English. A list of interviews is provided in the references; the detailed interview protocols are on file with the author. Besides empirical observations, clumsy floodplains are also legally analysed. The legal analysis refers to the period when the interviews have been conducted. Recent developments in legislation (water law was reformed at the beginning of 2010) are shortly described in the respective chapter. Regulations of the Federal Water Act after the reform of the Federal Flood Control Act in 2005 are indicated with 'WHG 2005', regulations of the recent reform of water law are referred to as 'WHG 2010'. The empirical and legal findings are then analysed with the help of Cultural Theory. With this theory, the social construction, which leads to clumsy floodplains, will be explained.

An approach for coping with extreme floods is presented in Chapter 2. After summarising contemporary discussions on flood protection, a new floodplain management based on retention is developed: Large Areas for Temporary Emergency Retention, LATER. The necessary engineering background is explained in this part of the book. Methods from economics as well as law and economics help to prove that LATER is the better technology in comparison to the contemporary levee-based flood protection. It will also explain why the clumsy floodplains are technologically locked-in into the levee technology, and the more efficient technology is not applied in practice.

For the implementation of LATER, land is needed. Making this land available is the concern of the last part of the book: how can we implement such a floodplain management in clumsy floodplains? Different proposals for land policies are derived from planning, economics, sociology and other disciplines: a competitive, a cooperative, a constitutional and a composed land policy are discussed. They are developed from the four rationalities described by Cultural Theory. According to this approach, the land policies should respond to the rationalities of the social construction. The proposals for land policies will therefore be discussed with this regard. Afterwards, a responsive land policy for LATER is presented and discussed.

This interdisciplinary research approach aims at a theoretical answer to the question how society could manage floodplains in the face of increasingly extreme floods. The answers proposed here, however, raise further questions. First attempts for implementing the here proposed idea in practice reveal particularities for realisation. But from responses in interviews and in expert discussions, it became obvious that the basic ideas, presented in this book, match practical requirements. This book provides no recipe for practitioners for a better floodplain management, but it provides a new approach to a better response to clumsy floodplains facing extreme floods.

List of Abbreviations

BauGB	*Baugesetzbuch* [German Federal Building Code]
BGB	*Bürgerliches Gesetzbuch* [German Civil Code]
DM	*Deutsche Mark*
EIA	Environmental Impact Assessment
EU	European Union
GG	*Grundgesetz* [Basic Law, German constitutional law]
IKSE	International Commission for the Protection of the Elbe River
IKSR	International Commission for the Protection of the Rhine
IPCC	International Panel on Climate Change
LATER	Large Areas for Temporary Emergency Retention
LAWA	*Länderarbeitsgemeinschaft Wasser* [Inter-state Working Group on Water]
ROG	*Raumordnungsgesetz* [German Regional Planning Act]
THW	*Technisches Hilfswerk* [German governmental organisation for disaster relief]
UBA	*Umweltbundesamt* [Federal Environmental Agency]
UVPG	*Gesetz über die Umweltverträglichkeitsprüfung* [Law on the Environmental Impact Assessment]
WHG	*Wasserhaushaltsgesetz* [German Federal Water Act]
ZÜRS	*Zonierungssystem für Überschwemmung, Rückstau und Starkregen* [information system on flooding, backwater and intense rain]

Chapter 1
Clumsy Floodplains

The Social Construction of Floodplains

Although more and more money is invested in flood protection (Loucks et al. 2008: 541, Strobl and Zunic 2006: 383), almost every year, extreme river floods cause enormous and increasing damage in floodplains (Munich Re Group 2003a: 14, Strobl and Zunic 2006: 411). The probability of extreme flood events increases as well in the future due to climate change (IPCC 2007: 8, UBA 2003: 1); in addition, our cities become more vulnerable (Cooley 2006, Fleischhauer 2004: 30–31). Usually water managers face this situation by building high and strong levees alongside the rivers. Such constructions, however, reduce the capacity of rivers to cope with the enormous water masses. Still, urban development takes place in these floodplains (e.g. Loucks et al. 2008: 541, UBA 2003: 148). Therefore, floodplains are contested land: places of extreme floods and places for urban development. The European Commission stated in 1999 that human activity supports these developments by:

> straightening of rivers, settlement of natural floodplains and land uses which accelerate water runoff in the rivers catchment areas (ESDP 1999: article 319).

Heinz Patt (2001: 12), however, emphasises that even without anthropogenic activity extreme floods can occur almost everywhere and that the less a flood is expected, the more it severe will be. Nonetheless, in the past and still today, urban development takes place in riparian landscapes and levees are necessary to protect the immobile values in the floodplains (Strobl and Zunic 2006: 391). As a consequence, space for the rivers is shrinking. Levees take space from the rivers. Along the River Elbe, for instance, 2,300,000,000 m³ of retention volume has been lost since the twelfth century due to levees. This is a reduction of 86.4 per cent of the original retention area from 6.172 km² to 838 km² (Engel et al. 2002: 3). This caused an increase of the flood levels up to 50 cm in Wittenberge (IKSE 2003: 24). Levees were heightened and strengthened. Eighty-five per cent of alluvial plains along the River Rhine diminished due to river constructions (City of Cologne 1996: 1–6). Similar developments happened at almost all big rivers in Germany. This lost retention volume is needed today to reduce flood risk in order to reduce flood damages (Strobl and Zunic 2006: 389). Space for the rivers is needed. Also internationally, 'land management is regarded as the most effective mitigation measures available' (Cooley 2006: 111). But giving space for the rivers is complex.

How should we use floodplains wisely in the future? In the aftermath of the 2002 flood in the Czech Republic and Germany, claims for withdrawing from urban land uses in the floodplains along the River Elbe emerged; but due to time, people lost sight of and forgot about the floods. Little by little, urban use in floodplains intensified again (Bahlburg 2005: 12). This pattern of human activity is characteristically in the floodplains: technical river constructions – especially strong and high levees – lead to a high discharge capacity of rivers; smaller flood events drain off without damage, which provokes additional urban developments in floodplains (Kron 2004: 9–12, Patt 2001: 2). In this manner, the contest between humans and the water continues: floodplains are contested land.

Floodplains are defined as potentially submergible riparian land. They encompass land between levees or embankments of a river as well as flat areas behind the levees, which could be inundated by an extreme flood or crevasse of levees. This understanding of floodplains encompasses the two categories of submergible land according to the German Federal Water Act (*Wasserhaushaltsgesetz*, WHG): *Überschwemmungsgebiete* (§ 76 WHG, formerly § 31b WHG) and *überschwemmungsgefährdete Gebiete* (§ 31c WHG 2005). *Überschwemmungsgebiete* are not sufficient for this research, because areas, which might be inundated by extreme floods or through crevasses are not covered by the definition in § 76 WHG respectively § 31b WHG 2005. *Überschwemmungsgefährdete Gebiete* according to § 31c WHG 2005 include such areas as well (whereas this category has been abandoned during the reform of water law in 2010). How is this potentially submergible land managed in the face of the contest between humans and the water?

Iteratively, humans behave in the same patterns of activity in floodplains. Most of the time, landowners are using floodplains profitably. They build houses, enjoy the riverside and just live close to the waterfront (Strobl and Zunic 2006: 391). When floods threaten their houses, water management provides sandbags and private relievers help to protect the houses. Then, water management agencies, supported by policymakers, build and improve levees in order to defend against floods. In consequence, planners perceive floodplains as secure and grant additional building permits to landowners, which accumulate additional value behind the levees. Barrie Needham describes the general pursuit for building in the Netherlands: 'A nation that lives, builds for its future' (2007a: 25). This is also the mood sustained by the German planning and property regime. So, floodplains are under pressure of urban development.

Levees ought to protect the immobile values in the floodplains. But levees protect only against smaller floods, up to thresholds. Extreme floods threaten the houses, economic enterprises, infrastructure and facilities in floodplains even behind levees. Additionally, levees capture the river and accelerate the discharge, which leads to higher water levels. Finally, flood risk increases, more flood damages result. The risk increase through levees and value accumulation behind the levees is more or less obviously observable along every river in Middle Europe. The risk increase results from prevalent patterns of human activity. It is in this manner not an environmental constraint but a social construction. At least in Germany, these clumsy patterns of human activity are apparent, as I will describe in the following section.

Patterns of Human Activity

A water manager at the flood protection agency of Saxony-Anhalt, concluded after the 2002 flood:

> *Bewusstseinsmäßig ist Einiges falsch gelaufen 2002. Einige haben sich einfach saniert. Die haben teilweise ein neues Haus bekommen. Die wollten zwar alle direkt danach wegziehen, aber jetzt ist alles neu und schön, die haben einen Deich und fühlen sich sicher. Wenn ein Schaden eintritt, dann sind wir verantwortlich.*
> [According to risk mentality, things went wrong in 2002. Homeowners wanted to move away after the flood, but then most of them renewed their homes; some even got a new house. Now everything is new and nice, they feel safe behind the improved levees. When a new flood event occurs, we, the water management agencies, are responsible.] (Water manager in Saxony-Anhalt in an interview in 2008)

These patterns of human activity persist and yield increasing flood risk: in the future all rivers will be captured by strong and high levees and intensively urbanised floodplains (IKSE 2003: 3). Although policymakers provide a lot of money for flood protection and set up directives to make space for the rivers (this policy was first set up in the Netherlands in the 1990s) (Greiving 2006: 73), municipal planning agencies grant building permissions in floodplains.

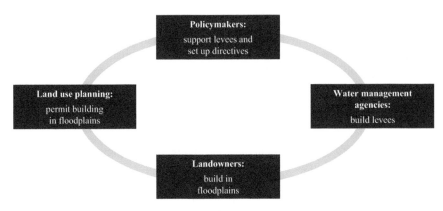

Figure 1.1   Patterns of human activity

Also in other European countries, planners struggle with providing space for the rivers, as Karen Potter concludes about the British system:

> The practice of floodplain restoration remains in its infancy and is not keeping apace with the policy rhetoric. (Potter 2008: 345)

Even extreme floods do not prevent landowners from building close to the rivers. Landowners and land use planners rely on levees and think levees can provide security against flooding. Finally, four stakeholders are essential for the described clumsy patterns of human activity in floodplains (see Figure 1.1): policymakers, land use planners, landowners, and water management agencies.

The activities of these four stakeholders correspond to and depend on each other, depicted in Figure 1.1 with the grey line. The diagram can be read counterclockwise as the following: landowners are not allowed to build in floodplains without building permission by the land use planners. Since landowners accumulate immobile values in floodplains, water managers must build levees. Values behind levees justify levees, but levees can only be realised if policymakers provide money for this purpose. Local planners can only permit building in floodplains as long as policy gives them some scope for local planning decisions. These social relations between the four essential stakeholders persist. It hinders concepts like space for the rivers.

Die Rheinpfalz, a German regional newspaper, wrote in July 2007:

> *Der Hochwasserschutz ist ins Wasser gefallen – An der Elbe sind fünf Jahre nach der Jahrhundertflut Deiche erhöht – Forderungen nach mehr Raum für den Fluß bleiben unerfüllt.* [Five years after the 2002 flood, the claim for more space for the rivers is not implemented, but levees have been heightened] (*Die Rheinpfalz*, 29 June 2007).

The policy for space for the rivers fails in Germany. Following the reasoning above, space for the rivers is not only a technological task, but also a social task. Since some years, water professionals explain their problems and solutions to other stakeholders and start to regard water management not simply as a matter of technical concerns (Lane 2006: xiii–xiv). Indeed, floodplain management is a misnomer. It is not the floodplain, but the people that must be managed (Bollens et al. 1988: 311). But the four stakeholders are entrenched in their patterns of activity. Why do stakeholders not change the situation and reduce risk by providing space for the rivers?

Timothy Moss and Jochen Monstadt (2008) explain the diffident mobilisation of space for the rivers with the high complexity of such projects. Policymakers have to cope with this complexity. Recent scientific publications often identify complexity as a problem of inter-institutional coordination, in particular between regional planning, municipal planning and sectoral planning (Düsterdiek 2001: 1201–2, Haupter and Heiland 2002, Stüer 2004: 416, UBA 2003: 149). Many authors emphasise regional planning as most important for coping with this complexity, and explain how regional planning should coordinate all relevant issues of flood protection in a catchment (Greiving 2003, Haupter et al. 2007: 518, Moss and Monstadt 2008, Seher 2004). Others claim for informal institutions (Zehetmair et al. 2008), or want the European Union to govern flood risk management (Arellano et al. 2007: 466–7).

In sum, these approaches are based on 'a widespread belief', consisting of 'three mental shortcuts': first, it is assumed that environmental management is a problem of coordination; second, consultation solves lacks of coordination; third, 'consultation is inseparable from consensus' (Billé 2008: 77).

> The consensual search for coordination via consultation underestimates the real antagonisms that exist between 'uncoordinated' stakeholders and uses, the differences of interests and of representations (Billé 2008: 78).

This antagonism has to be described to analyse the complexity of floodplains management. It cannot be explained with surveying facts, weighing up probable costs and benefits of various possible interventions, and moving from there to the right answer (Ellis and Thompson 1997: 208). Rational choice theory, where people act rationally in order to maximise their utility (Cooter and Ulen 2004: 15), cannot explain the whole story of the floodplains. We continue to observe the opposite of 'wise' and 'rational' floodplain management (Loucks et al. 2008: 541).

Cultural Theorists support the position that such a 'Newtonian policymaking' approach is appropriate for noncomplex systems; but it fails in complex systems (Ellis and Thompson 1997: 208). Social systems, however, are rather complex like ecosystems (Ellis and Thompson 1997: 209). In such social systems, cost–benefit analyses, probabilistic risk assessment and general equilibrium modelling will not be useful tools in designing and redesigning institutions. Moreover, the contending social framework of problem and solution between different audiences must be identified (Ellis and Thompson 1997: 209). Like in coastal zone management, Raphael Billé writes, floodplain management is not managed by a manager, but it is rather 'a process without a pilot' (Billé 2008: 79). So, many different stakeholders contribute to the social construction of floodplains. This frames floodplains as complex social systems. Identifying the social system requires an analysis of the contending perceptions of different stakeholders in floodplains.

Perceiving Floodplains

If you jump into the water, you will get wet. Regarding floodplains as social systems does not neglect this fact. Floodplains are certain pieces of riparian land. You can measure their size, determine the soil conditions, or identify land uses by remote sensing. Following these statements, there is nothing social about floodplains. Really? What are typical attributes that an inhabitant of the floodplain would assign to these landscapes? How would water management engineers declare floodplains? Insurance officers, planners, politicians, foreigners, tourists, disaster management workers, farmers, captains … different social actors assign different attributes to floodplains – depending on their 'condom of monorationality' (Davy 2008: 304). What are these condoms? These condoms represent their rationalities. They are like filters in people's heads, which protect them from being raped by the plural rationalities (Davy 2008, based on Simmel 1903). For example, as a tourist, it would not be suitable to regard floodplains through the eyes of a farmer. For the one, floodplains serve as places for recreation, for the other the same place is a working place. The condoms – the rationalities – protect individuals from being overwhelmed by the different rationalities. Floodplains are socially constructed through different rationalities – through different perceptions and filters for reality. This does not necessarily mean that everything that happens in floodplains can exhaustively be explained by social constructions:

> Rather, ideas of nature are plastic; they can be squeezed into different configurations but, at the same time, there are some limits (Thompson et al. 1990: 25).

Rationalities influence patterns of human activity. Indeed, rationalities are often not apparent; rather they are hidden as different cultures, ways of life, biases and perceptions (Thompson et al. 1990). To uncover the rationalities, which underlie the social construction of floodplains and frame the activities of the stakeholders, the four most important stakeholders have been interviewed. The discussions with policymakers (officers at ministries and regional planners), local land use planners (and mayors), water managers and randomly selected landowners in flood-prone areas in Cologne-Rodenkirchen, Biederitz near Magdeburg, Dessau-Waldersee, Bitterfeld and Torgau uncovered different perceptions of floodplains in different phases. The qualitative empirical research was conducted in semi-structured narrative interviews of approximately 15 to 40 minutes. The aim was to learn more about the perceptions of the stakeholders. The following four perceptions have been identified out of the interviews, observation, analysis of literature and other documentation according to flood events (e.g. newspaper, DVDs, publications, radio and TV reports): floodplains are profitable, floodplains are dangerous, floodplains are controllable, and floodplains are quite as inconspicuous as any other terrain. These four perceptions of floodplains differ essentially.

Profitable floodplains Floodplains attract a number of land uses. Large settlements are likely located along rivers (Cooley 2006: 91, Petrow et al. 2006: 717). Since the beginning of the explosive growth of populations in the nineteenth century, and due to the industrial revolution, floodplains became more and more attractive and increased in value; engineers and planners protected these areas with levees (Strobl and Zunic 2006: 384). New developments took place in flood-prone areas (Bahlburg 2005: 12, Greiving 2006: 73). Why do people live in floodplains? Landowners reacted to this question in an astonished way: '*Warum wir hier leben? Schauen Sie sich doch mal um! Es ist herrlich hier!*' [Why we are living here? Look at the beautiful landscape! It is great here!], or: '*ich kann von meiner Terrasse aus den Rhein sehen*' [I can see the River Rhine from my terrace].

Indeed, the reasons for settling in floodplains vary, but often the beautiful nature and the scenic landscape were emphasised. For most people interviewed, flooding was no topic for their individual allocation decision. Often landowners were not aware of the possibility of flooding when they bought or built their homes on floodplains. Some were warned by experienced neighbours (e.g. in Biederitz near Magdeburg), nevertheless people moved into the riparian landscapes. Municipalities rarely inform landowners about the risk of flooding. The interviewed landowners confirmed that only few moved away after a flood, and most would even stay after further extreme floods: '*Hier ist unser Geld, hier ist unsere Anlage*' [Here is our money, here is our investment]. People still demand riparian building sites; building in floodplains is a profitable business (Bollens et al. 1988: 312).

> Floodplains provide fertile farmland, recourses, for economic development and drinking water, and they act as corridors for transport (Petrow et al. 2006: 717).

High quality residential, commercial and industrial areas have been developed in the past and will still be developed in the future. Planners, investors, architects, engineers, building companies – a whole industry thrives from building in floodplains. Interviews with land use planners confirmed that spatial planning supports this industry by granting building permits, mayors are lobbying building in floodplains. At the local level, economic interests for building in floodplains often drowns obligations of water managers (Bahlburg 2005: 12). In particular, local enterprises located in flood-prone areas need additional land in floodplains for extensions of existing facilities with the argument of workplaces (Greiving 2006: 73). Since local economies often rely on such enterprises, it is likely that municipalities will fulfil their demands.

Landowners feel relatively safe behind high and strong levees. The majority of landowners affirmed that they could cope with further floods: clean up the mess, rebuild and stay. A usual response was that people who live in areas at risk have to arrange themselves with the threat. In conclusion, landowners who already live in floodplains stay and bear the risk, because they want to enjoy their property and rate this as an important value.

According to new developments in floodplains, interviewed water managers complained about municipal planners: they would often not believe in statistical risk calculations. They do not want to hear about restrictions, but rather opportunities for urban development. At the municipal level, flood risk maps are threats for new urban developments. Water managers explained that land use planners at the municipal level justify urban developments in floodplains with personal experience of a few years when no flood occurred. Regional planners affirm these statements: if no flood occurred for some years, municipal planners perceive floodplains as attractive land for building. Restrictions from water management or regional planners are not welcome at this time. Regional planning is almost unable to restrain building in floodplains, as it came out in the empirical study. Indeed, questioned local land use planners emphasised that building in the floodplain is essential for the economic and social future of municipalities. Especially for bigger towns it would be necessary to provide attractive building sites for young families, because those towns need inhabitants and they have to restrict urban sprawl by designating building land on their territory. Like landowners, municipal planners rely on levees. Planners at the local level are not willing to restrict building in levee-protected floodplains.

> *Wir verlassen uns auf die Sicherheitszusagen der Wasserwirtschaft.* [Town planners rely on skills of water management] (Town planner in an interview in 2008).

Even in areas in very risky regions, like in some parts of the Netherlands, the reliance on technical solutions is high (Roth and Warner 2007: 519). So, this belief in the controllability of floodplains is not a particularly German phenomenon. Even after extreme floods, building in floodplains persists, for example in Dessau-Waldersee in Eastern Germany, which was inundated completely in 2002; people still did not build risk-adapted buildings.

> *Wir haben nach 2002 hier alles wieder aufgebaut, wir fühlen uns jetzt sicher.* [After 2002, we have rebuilt everything, and we feel safe now] (Mayor of a small municipality in an interview in 2008).

In particular, landowners and municipalities seem to perceive riparian landscapes predominantly as profitable areas when no flood occurred for some years. Landowners gain happiness and welfare; municipalities seem to have a key interest in accumulating satisfied landowners behind the levees. Thus, they grant building permits. Indeed, floodplains are profitable!

Dangerous floodplains Sometimes floodplains are not perceived as profitable but as rather dangerous areas. In the years since 1993, an extreme flood event has occurred almost every year in Germany. Each time, the inundations caused enormous harm. When a flood occurs, disaster control measures start. With enormous effort, each square metre of floodplain is defended against the water. Sandbags, sealing foil and a lot of manpower are mobilised for that reason. After the water has receded, cleanup and reconstruction measures start. This is the normal procedure of flood risk management (DKKV 2003: 10). Flood response is 'largely based on crisis management' (Cooley 2006: 92). Very often, these emergency measures and reconstructions are supported by charitable donations and voluntary aid (Helbig 2003: 118–25). Indeed, disasters initiate egalitarian patterns of activity. Landowners and water managers confirmed that victims of a flood usually commonly defend the threat:

> *Beim Hochwasser gilt das Solidarprinzip.* [When a flood occurs, solidarity is essential] (Town planner of an affected city in an interview in 2008).

Most landowners emphasise the socialising character of dangerous flood events:

> *Das Hochwasser hat uns alle hier zusammengeschweißt.* [The flood bound us to each other] (Landowner at the Elbe in an interview in 2008).

Floods seem to strengthen neighbourhoods and bring people together. Many landowners were surprised about the solidarity during the 2002 flood at the River Elbe. Not only victims but also people from all over Germany supported the disaster relief measures. Some landowners established civil initiatives or would participate in such an organisation in order to achieve better flood protection (Patt 2001: 3). They are convinced that only with a strong community do they have a chance to defend against the water. Often, the state steps in: after the 2002 flood, the German federal government passed a law on solidarity with flood victims (*Flutopfersolidaritätsgesetz*). It made 7,100,000,000 Euro available for reconstruction and relief measures (German Bundestag 2002). Also in 2002, the European Commission established the European Union Solidarity Fund, which enables the Union to grant rapid financial assistance in case of emergencies through major disasters (European Union 2007: 28).

Asking landowners and water managers shows that floodplains can sometimes be dangerous terrain. In particular, the property of landowners is threatened. In this situation, municipal land use planners are less dominant in determining the management of floodplains. Rather, disaster relief is the apparent management approach. Floodplains are then not profitable but rather dangerous land.

Controllable floodplains   In the aftermath of floods, victims usually request higher and stronger levees, explained a water manager of Saxony-Anhalt in an interview in 2008. Almost all interviewed landowners agreed that flood protection is a governmental task. Whereas landowners interviewed in Cologne-Rodenkirchen emphasised their additional self-reliance, landowners along the River Elbe rely more on the state. In Dessau-Waldersee, one of the heavily affected areas during the flood of 2002, people formed citizens' initiatives and demonstrated for high and strong levees (Helbig 2003: 129–30). Indeed, landowners are claiming for better protection in the aftermath of floods, but when measures start, most civil initiatives abandon their mission or become inactive.

As a first response to the civil initiatives, politicians often promise quick actions. Water management agencies do then have to implement these promises – no matter whether the actions, e.g. strengthen a levee, fit in the water management concept, are meaningful or not – criticised a water manager in Saxony-Anhalt. Policymakers want to demonstrate full control of the situation. Water managers affirm:

> *Jeder im Wasserbereich sagt, dass ein Hochwasser alle paar Jahre nötig ist, um die Erinnerung wach zu halten.* [Everyone in the water sector ascertains that every few years a flood is necessary to keep risk awareness conscious] (Water manager in Saxony-Anhalt in an interview in 2008).

The water managers in Saxony-Anhalt complained that without a doubt, a lot of money was available for flood protection measures after the flood event in 2002, but it had to be spent too rapidly. After some time, when the public interest disappeared, policymakers designate the unused money for other purposes (water manager in Saxony-Anhalt in an interview in 2008). This implies a high pressure for implementation. Water managers ascertained during the interviews the difficulties with implementing space for the rivers and that plans for restoring floodplains are long-term projects, because they intervene in the property rights of landowners of the floodplains (water manager in Saxony-Anhalt in an interview in 2008). Improving the levees structurally was an immediate opportunity for water managers after 2002. Thus, politicians provided money for levees. High-tech levees along the whole floodplain are still the dominant response to flood risk.

> For centuries, river management in Germany, as in many other countries, has been powerfully shaped by the drive to control floods by 'hard engineering' methods (Moss and Monstadt 2008: 70, Johnson and Priest 2008: 514).

Not only Germany, but other European countries as well are entrenched in technological responses to floods. In particular, the UK has its traditionally favoured technological solutions (Brown and Damery 2002: 412). These patterns of activity after floods worsen the situation for downstream parties, but it provides some security for local landowners (water manager of Saxony-Anhalt in an interview in 2004).

Many authors confirm that flood protection is often 'a "command and control" approach, based on engineering works' (Roth and Warner 2007: 519). Local politicians have an interest in having their floodplains protected against a centennial flood event, which means in their terminology 'flood-risk free' (Loucks et al. 2008: 542). In this manner, local politicians promote levees and it appears to landowners that flood risk is 'well in hand' (Loucks et al. 2008: 542).

In addition to this technical flood response, often policy directives are set up to improve precautionary flood protection (Moss and Monstadt 2008: 65–8). Already in 1996, policymakers intended to reduce building in floodplains (DKKV 2003: 10). The guideline for flood protection of the future from 1995 by the Interstate Working Group on Water is an important policy document (*Länderarbeitsgemeinschaft Wasser*, LAWA) (LAWA 1995). The five-point programme for improving the precautionary flood protection by the Federal Government of Germany (German Government 2005) is a recent policy response in Germany, whereas the major reform of the water law in 2010 cannot be seen as a direct reaction to a flood. After floods, many policy documents ascertain the ability to control the flood risk.

These legislative attempts ought to demonstrate: floodplains are controllable! Claims for protection measures would be irrational if policymakers did not believe in the controllability of the floodplains. So, usually after a flood the public investments are high. Water management agencies build technical constructions in order to control the rivers. Policymakers pass legislations and set up directives to demonstrate the ability to cope with the situation. Obviously, floodplains are controllable.

Inconspicuous Floodplains

> A flood event is almost forgotten after only a few years (Petrow et al. 2006: 720).

Flood protection is only temporarily a political issue. Once legislation has passed, other topics appear on the political agenda. Terrorism, bird or swine flu, environmental issues or economic crises are the topics politicians pay attention to. Laws and regulations for improved flood protection remain to be implemented on the executive local level. The Federal Environmental Agency concluded in 2003 that decision-making at the local level follows a distinct logic: the message that implementation of space for the rivers is an important planning issue would not yet have been heard at the local level (UBA 2003: 148). Water managers are complaining about a lack of risk awareness from landowners and local planners after the mess of floods has been cleaned up. Water managers perceive it as an important task to convince municipal planners and landowners about the danger of living in floodplains (water manager in Saxony-Anhalt in an interview in 2008). Water management agencies build levees, although they know that strengthened and heightened levees in fact heighten discharge levels and only pretend security. This leads to increasing vulnerability and more potential damages. Building levees and accumulation of immobile values behind them is a *fortschreitende Entwicklung* [a continuing development] (water managers in Saxony-Anhalt in an interview in 2008). Some landowners claim that municipalities should not allow building in the floodplain any longer. Rather, municipal planners should take responsibility for their plans; politicians should not support urban development in riparian landscapes any longer. These statements from landowners demonstrate how critical they are about the developments in floodplains. One landowner pointed out that some houses should probably not have been built in floodplains, but since municipalities granted building permits and thus these houses exist, flood protection should be provided by the state (landowner near Magdeburg 2008). This opinion represents the majority of the interviewed landowners, but no one amongst the questioned landowners is willing to protest against additional building in floodplains, even if it affects their own protection level (for example if an area upstream is developed). Municipal planners believe that politics must regulate building in floodplains. Additionally, they hope landowners are sensible and take care of their risk (land use planner in an interview in 2008). Policymakers have another point of view: mayors are self-responsible to adapt old zoning plans to the increasing flood risk. According to new developments, policymakers can, to some extent, influence regional planning, but constitutional local self-government confines this opportunity. Particularly for existing settlements, regional planning has almost no instruments for intervention (land use planner in an interview in 2008). In the end, every stakeholder is complaining about the others. Nonetheless, nobody takes action; the stakeholders emphasise general responsibilities of the others.

In this situation, floodplains are inconspicuous terrain. Areas, which were designated for building before a flood event, will be realised after the water is abandoned (water manager in Saxony-Anhalt in an interview in 2008). For example in the small village Gübs, which was flooded in 2002, 25 new single-family houses will be built. The particular sites were designated long before the flood at the Elbe happened, so they can be realised by now (without risk adaptation).

> *Der Bebauungsplan wurde bereits 1992 genehmigt.* [The zoning plan was approved in 1992] (Mayor of a small municipality in an interview in 2008).

The mayor was convinced that landowners should be able to build in the floodplains, because they provide nice locations and cheap land. This is not an irrational reaction. Why should one extreme flood event paralyse urban development in every floodplain? The flood issue is just one of many relevant topics for spatial planning, not more or not less important than others, confirmed an officer at the Ministry for Building and Transport of Saxony-Anhalt, in an interview in 2008. Why bother with inconspicuous land? Each of the stakeholders, water managers, landowners, mayors and municipal planners, and policymakers complained about the lack of risk awareness, but nobody pays specific attention to this particular topic. Floodplains are inconspicuous!

Persistent Clumsiness

Which lessons have stakeholders learned from past floods? Even the series of extreme flood events in the past 15 years could not change human activities in the floodplains: still, values are going to be accumulated, destroyed, rebuilt and extended in riparian landscapes. This is clumsy. The Federal Environmental Agency (*Umweltbundesamt*, UBA) criticised already in 1998 that after the 1997 Oder flood new buildings were built again behind the recently reconstructed levees (UBA 1998: 97). After 2002, financial donations and governmental aid supported reconstructions in the affected regions along the River Elbe. Interviewed landowners confirmed that, in total, the financial support was much more than sufficient. As an effect, landowners were able to rebuild their homes, and sometimes even to improve them, but most had not adapted to the risk – although such techniques are available. Some even expect similar help and compensation after the next flood, as these statements of interviewed landowners emphasise:

> *Ich hab kein Minus gemacht.* [I have not made a loss].

> *Das Hochwasser hat sogar einen positiven Effekt gehabt, schauen Sie sich um, alles neu hier die Straßen, die Häuser, das wurde alles bezahlt!* [The flood had even produced a positive effect, look around here, everything is new: the streets, the houses, everything was paid].

> *Wenn nochmal das Hochwasser kommt, lass ich die Firmen kommen, um alles zu renovieren.* [Next time, I will call professional companies to clean up the mess] (Landowners in Dessau Waldersee in an interview in 2008).

These lessons, learnt from flood events, do not prevent harm. On the contrary, clumsy patterns of human activity persist. Are people not willing to change their behaviour?

A persistent social construction   The patterns of human activity in the floodplains are driven by certain perceptions of the floodplains. These perceptions represent a persistent social construction of floodplains by essentially four stakeholders: landowners, land use planners, water managers and policymakers (see Figure 1.2).

The actions of stakeholders are driven by one of the four described perceptions. In the model, each pair of stakeholders shares a certain perception in a particular phase of the floodplains. By acting in accordance with these perceptions, activities take place that sustain contemporary management of floodplains: a reiterative procedure of building in, defending, protecting and ignoring floodplains.

The perceptions are not 'personal perceptions', assigned to individual stakeholders (although an agency for instance is not an 'individual', its activity is individual in relation to the other stakeholders). Rather, activities of individual stakeholders are determined by time and place. The perceptions are the rational

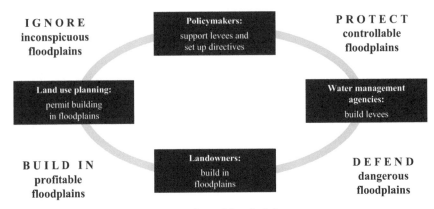

Figure 1.2   The social construction of floodplains

basis for the stakeholders 'at time and in places'. Michael Thompson (2008b: 19) emphasises that there is nothing 'individual' about the perceptions (indeed he speaks about rationalities, which will be elaborated later). The physical individuals in the social construction are rather social 'dividuals'. 'Dividual' does not refer to the fact that each stakeholder is indeed composed of many people, but rather the social embedment is dividual. In this manner, the whole socials construction determines patterns of activity before, during and after floods.

Land use planners and landowners build in floodplains in dry phases. Land use planners and landowners share the perception that floodplains are profitable. Probably, other stakeholders, inside or outside of the depicted social construction, perceive floodplains similarly – insurance officers for example might profit a lot from floodplains, but planners and landowners are essential for the accumulation of immobile values in floodplains. Without the permission of land use planners, landowners could not act in that way, without the demand of landowners, land use planners would not designate these areas for building.

Landowners and water managers defend floodplains in case of floods. They share the belief that floodplains can be dangerous. The destructive forces of the water affects landowners. Since water managers perceive floodplains as dangerous terrain, they help landowners with sandbags and sealing foil to defend the water. As these two stakeholders share the perception of dangerous floodplains, disaster management starts and chains of voluntary workers erect walls of sandbags on top of the levees.

Water managers and policymakers protect the floodplains after a flood. After the disaster is relieved, water managers believe in the controllability of floods. If floodplains were only dangerous, water managers would never agree with building a levee. Policymakers and water managers believe that floodplains are not dangerous but rather controllable. Politicians provide money and political support for water managers, who have the expertise for building strong and high levees.

Policymakers and land use planners ignore floodplains most of the time. If that would be the whole story, no new building would emerge in floodplains. But policymakers and land use planners are essential players for granting building permits in the floodplains. Both agree – without expressing that thought explicitly – that floodplains are inconspicuous. No special restrictions must be made for local planners. No specific planning tools are required for this inconspicuous piece of land – the floodplains.

Figure 2.1 (Chapter 2: 49) shows the shared beliefs and the patterns of activity of the four stakeholders. In the end, clumsy floodplains are managed by essentially four stakeholders in an iterative loop of building in, defending, controlling and ignoring floodplains.

Phases of floodplains Finally, something does not fit. As explained, patterns of human activity of the four stakeholders are socially constructed by distinct perceptions. Each of these perceptions claims to depict the floodplains. Each claims to represent the truth, claims to be rational. Remarkably, each response is right, but in sum, they do not fit together. Floodplains cannot simultaneously be profitable, dangerous, controllable and inconspicuous. But the array of these perceptions leads to a clumsy result: risk increases. How are they arrayed? Flood risk management is a permanent process of disaster, reaction, recreation and improvement (Grünewald 2005: 15). Indeed the perceptions are arrayed in phases. Perceptions change due to time (Zehetmair et al. 2008: 206–7). They change when it becomes obvious that a situation is not explainable by the contemporary dominant perception. This means, there are phases where floodplains are profitable, phases where they are dangerous, controllable and inconspicuous. Figure 1.3 illustrates these phases. Each phase can be measured economically: the monetary investments in building activity, the damage statistics, and investments in levees indicate the economic effects of the clumsy floodplains. The diagram shows no absolute relationship to the monetary aspects

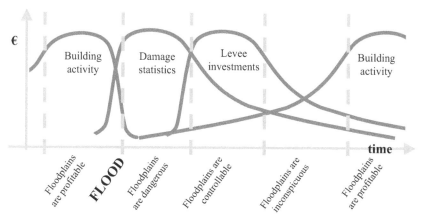

Figure 1.3   Phases of perceptions in floodplains

of these phases; rather it schematically shows the ups and downs of each phase in relation to the ups and downs of the other phases due to time. This also implies that in each phase distinct stakeholders' perceptions dominate what happens in floodplains.

The phase of profit: the phase where floodplains are profitable is a phase of high building activity. Big building projects at the rivers like the *Rheinauhafen* in Cologne or the *Rheingallerie* in Ludwigshafen at the Rhine take place quite a while after an extreme flood (for the Rhine, the last flood was 15 years before the two projects were launched). The investments in the floodplain are an index for the profitability of the floodplain. A flood event, which threatens the values in the floodplain, causes a change in the dominant perception. Land use plans for new residential areas in riparian landscapes, or plans for the allocations of commercial centres near rivers, are not welcome in this phase. Rather, landowners and land use planners dominate the scene.

The phase of danger: damage statistics represent the dangerous floodplains. Remarkably, even damage statistics vary due to different phases (Schwarze and Wagner 2004: 154). The Munich Re Group analysed official damage statistics during a long period: during the catastrophes, the estimated damage is less high; later, the damage forecast increases to high extents – the damage will then often be overrated. Floodplains are perceived as dangerous terrains then. The Munich Re Group assumes that, in this phase, the statistics are used as political instruments to justify costs for disaster management (especially because the governmental parties often profit from catastrophes, for example, in the elections of the German parliament in 2002, when Chancellor Schröder gathered votes against his opponent Stoiber as a consequence of the flood event). After the public interest abates, a long time after the catastrophe, the statistics draw lower damage – floodplains become inconspicuous (Munich Re Group 2003b: 27).

The phase of controllability: when damage statistics are at a moderate but still relatively high level, investments in levees increase: 1.232 km of levees exist along the River Elbe in Germany (except the tidal-influenced part of the Elbe). In the 11 years before the flood of 2002, namely since 1991, 197 km of these levees had been improved at a cost of 154,000,000 Euro. Until 2014, approximately three times as much, namely 548 km, is being improved with estimated costs of 591,000,000 Euro. This means that this flood event caused a triplication of velocity and an increase of costs (IKSE 2003: 77). After the flood events at the River Rhine in 1993 and 1995, the German state Hessen invested in 1997 approximately 24,000,000 DM (that is about 12,000,000 Euro) in levees along the River Main and the River Rhine (Corell 1996: 246). Typically, after extreme floods investments in levees and technical infrastructure rise. These statistics strengthen the argument that a phase exists where policymakers and water managers perceive floodplains as controllable terrain.

The inconspicuous phase: inconspicuous floodplains cannot be measured in money, because they are also in this matter inconspicuous. This phase is in between the phases after disaster – when cleanup measures, reconstructions, etc. take place.

In sum, this sequence of phases of perceptions contributes to a persistent social construction of floodplains, which sustains patterns of human activity and finally aggravates flood risk. The social construction describes a clumsy practice of treating floodplains. Why does this clumsiness persist?

An institution is a 'significant practice, relationship, or organisation in a society or culture', following the definition of the Merriam Webster Online Dictionary (www.merriam-webster.com). Michael Thompson suggests regarding any 'non-randomness' of behaviour or of belief as an institution (2008b: VI). In this meaning, the social construction yields an institution:

> [Institutions] consist of formal rules such as laws and regulation and informal constraints such as conventions, norms, and traditions (Van Rij 2008: 6).

Marco Verweij, Michael Thompson and Christoph Engel emphasise that institutions are socially constructed by different rationalities. The institutional settings result from alternative combinations of the rationalities (2006b: 215). It does not help to go deep into definitions of institutions or add a contribution to discussions about institutionalism, nor would it be relevant to discuss whether the social construction is an institution or not. But the statements of Michael Thompson, Evelien van Rij, Marco Verweij and Christoph Engel help to structure the analyses of the social construction of floodplains.

What are the 'formal rules' and the 'informal constraints' of the social construction? Which combinations of rationalities frame the setting? The formal rules will be identified by analysing the relevant law. Subject of this analysis will be the German law in the following chapter. The informal constraints refer to the underlying ways of thinking (Van Rij 2008: 6). It affects the rationalities pursued by stakeholders in certain phases. Let us start with the analysis of the legal framing of the social construction and proceed afterwards with the rationalities.

Planning, Law and Property Rights for the Social Construction

> The German system for land use planning and water supply is '... highly complex, fragmented and decentralized' (Moss and Monstadt 2008: 68).

The legal framing of the activities of landowners, water management agencies, policymakers and land use planners will be analysed in the following sections. Other stakeholders like insurance firms, regional planners, sectoral planning agencies or experts might also be involved, but these four are essential for the described social construction since these four act out certain rationalities in certain phases.

Landowners

Under which conditions may landowners build in floodplains? In Germany, several laws must be consulted to answer this question. The constitutional Basic

Law of Germany (*Grundgesetz*, GG) provides the constitutional aspects for using private properties. Legal framework for land use in general can be found in the Federal Building Code (*Baugesetzbuch*, BauGB). The Federal Water Act (*Wasserhaushaltsgesetz*, WHG) determines regulations for building in floodplains, whereas the Flood Control Act (*Gesetz zur Verbesserung des vorbeugenden Hochwasserschutzes*), that reformed the WHG in 2005, also plays an important role. These laws also provide an answer to the specific responsibilities and liabilities of landowners with respect to flood protection.

Building in floodplains   Article 14 I GG guarantees private property. Landowners are entitled to use their property freely and keep the gain of using it (Davy 2006: 56). In addition, § 903 German Civil Code (*Bürgerliches Gesetzbuch*, BGB) grants free usage of property as long as rights of others are not affected. Restrictions for the use of land are determined by the law (article 14 I GG). Property entails obligations; its use shall also serve for the public (article 14 II GG). However, this obligation does not mean that private landowners should drown out individual interest by public interest; individual landowners should not be disadvantaged unilaterally (Davy 2006: 66–7). Building houses and accumulating values on property is covered by the constitution as long as these actions comply with the legislation.

To what extent landowners are entitled to build on their land is locationally specific (*situationsgebunden*) (Needham 2007c: 184). Spatial planning specifies the situational relation of property in accordance with the Federal Building Code (Federal Supreme Court of Justice, in NJW 1957: 538–9). The German Building Code regulates in § 30–37 BauGB under which conditions building is permitted in Germany. Three cases are distinguished: first, building is permitted if the project is in accordance with a land use plan under the conditions of § 30 BauGB; second, building is permitted in areas, which form a cohesive settlement, under the conditions of § 34 BauGB; in all other cases, building projects for specific functions are permitted, and strict conditions have to be met (§ 35 BauGB). One of these conditions is that water management and flood protection must not be affected (§ 35 III no. 6). However, the German Building Code defines no rule that prohibits building on sites, which could be flooded. In other words, it is not illegal to expose values to floods.

The Federal Water Act (*Wasserhaushaltsgesetz*, WHG) imposes restrictions for building in flood-prone areas. In 2005, the Federal Flood Control Act implemented important additional clauses in the Federal Water Act. Due to the reform of federalism in 2006, the implementation of these regulations depends on the legal application in the federal states (*Länder*) (Moss and Monstadt 2008: 73). The 2010 reform of the Federal Water Act had also some implications for building in floodplains.

According to the Water Act after 2005, the federal states had to launch adequate laws to regulate land use in floodplains with respect to environmental aspects of using the floodplains, restoring floodplains, high water discharge and the prevention of damage (§ 31b II sentences 6–7 WHG version 2005, Moss and Monstadt 2008: 73). Since March 2010, the reformed Federal Water Act strengthened this regulation since water law is no longer just a framework–legislation, which has to be translated in the legislations of the federal states, but rather direct legislation of the federation (German Bundestag 2009). Since 2005, the WHG has remarkably framed the compensational aspect of these regulations: economic disadvantages for agricultural and forestal land uses due to these regulations shall be compensated (§ 31b II sentence 8 WHG 2005 respectively § 78 V sentence 2 WHG 2010). Originally, compensation for agricultural and forestal uses was introduced in the WHG in 1996. The legislator wanted flood protection not to be extended at the expense of farmers (German Bundestag 1996: 21). In 2005, the restrictions, but not the compensation, were extended to the settled areas (§ 31b II sentence 4 WHG 2005), whereas the 2010 version of the WHG is slightly more restrictive for agricultural land use of floodplains (§ 78 I WHG 2010). For disadvantages in settled areas due to the regulations, no compensation is provided. It is prescribed that such regulations are not an expropriation in the sense of article 14 III GG but a restriction in the sense of article 14 I sentence 2 GG (Cormann 2008: § 31b marginal 4). Finally, restrictions for settled areas might burden private landowners without compensation. In 2005, it remained to be seen how regulations in § 31b WHG 2005 would be transferred in the legislation in the federal states and how water authorities execute the rules. It was expected in 2005 that the 16 German federal states would implement these regulations differently in their federal water laws, as experiences with past reforms show (Breuer 2006: 615). However, before an evaluation could take place, the new Federal Water Act was launched in 2010 and regulates that water authorities have to decide on restriction under a set of rules, determined in § 78 WHG 2010. In the end, landowners might be restricted in their land use by the water authorities, but they need not be compensated for building restrictions.

In addition to the regulations that the federal states have to launch for floodplains, since 2005, the Federal Water Act contains a further clause according to building in floodplains. Landowners, who want to build in the floodplains in accordance with § 30, § 34 or § 35 BauGB, need an additional authorisation to receive a building permission (Cormann 2008: § 31b WHG marginal 32). Several conditions had to be matched for that additional authorisation: retention and flood discharge must not be affected, lost retention volume must be compensated physically (not financially), flood protection must not be affected and the constructions must be built risk-adapted to inundations (§ 31b IV sentence 3 WHG). In case of refusal, landowners are also not entitled to compensation (Breuer 2006: 622). In 2010, the legislator strengthened the restrictions for land uses in determined inundation zones. § 78 WHG 2010 (former § 31b WHG 2005) contains a list of prohibitions. Land users (and others) may not build walls or dams that might hinder water discharge in case of inundation. Disposing anything that could contaminate water or soil is prohibited, whereas it is remarkable that again exceptions are possible, if substances serve agricultural or forestal land uses. Changing the shape of surface (heightening or deepening) is also prohibited in formally determined inundation zones (§ 78 WHG 2010).

In 2010, another rule has been abandoned: after 2005, regulations for land use in floodplains were not applicable for all rivers, but only for rivers, or part of rivers, which have been designated by the agencies at the level of the federal states as areas where 'more than minor damage' is expected in case of flooding (§ 31b II sentence 1 WHG). Thus, rivers and parts of rivers could be excluded from the regulations. This would induce that particularly rural areas would have been excluded, because along rivers and parts of rivers in these areas, less damage can be expected. Proprietors of land in rural areas would therefore have been better off. The reform in 2010 implemented the Flood Directive of the European Union 2007/60/EG. It claims to prepare hazard maps and risk management plans for whole catchment areas (German Bundestag 2009). So, the recent reform had some crucial implications for land users in rural areas. Restrictions indeed affect building in floodplains, however it is remarkable that the legislator wanted to strengthen the regulations at the expense of farmers.

It can be concluded that building in floodplains has become increasingly difficult in Germany due to recent legislation. In 2005, no general prohibition has been introduced (Kotulla 2006: 135). In 2005, the Federal Water Act imposed some restrictions for the land use of this land, but these restrictions depend on the particular implementation in the federal states. The regulations were implemented quite differently in the individual federal states (Kotulla 2006: 135). Until 2010, building was still possible in floodplains. The new regulation in 2010 includes for the first time an explicit prohibition for building in formally declared floodplains. Although, exceptions are still possible under certain conditions, this new regulation is seen as a great success (Rolfsen 2009: 770). Restrictions for building in these zones do not have to be compensated. However, in areas behind levees, which might be affected by a flood if a levee breaks, building is not prohibited. Still, values can be accumulated and exposed to extreme floods. Furthermore, there exist already many settlements in the floodplains – either in front of or behind the levees. The regulations for building permission and restriction do not address such existing developments. So, the question emerges what are the responsibilities and liabilities for these sites?

Responsibilities and liabilities   What is the legal position of landowners in case of flooding? Inundations cause damage. Landowners suffer financial loss. The state is not generally liable. Who is liable for the losses? Often the state steps in – like in 2002, when chancellor Schröder postponed a tax reform to relieve flood-affected landowners (German Bundestag 2002). However, the state is not formally liable for individual damages (Reinhardt 2004: 420). Individual landowners or land users may not derive any sueable right for particular flood protection measures by the state, the municipalities or the water management agencies (Breuer 2006: 618, Schwendner in Sieder et al. 2007: § 28 WHG marginal 12d). Only in cases of culpable official liability, or in cases of governmental action in the sense of scarification or expropriation, may landowners receive compensation (Reinhardt 2004: 420). Private landowners are not only entitled to gather the advantages of property, but also the disadvantages (Davy 2006: 63). Landowners may not formally claim for the socialisation of their financial losses. They may not generally blame anyone for their losses, except if their right of physical integrity is affected (Knopp in Sieder et al. 2007: preface, marginal 19a).

In 2005, a general clause for responsibility for flood protection was implemented in the Federal Water Act. Everybody who might be affected by inundations is obliged to prevent damage (§ 31a II WHG 2005, § 5 II WHG 2010). This implies that flood protection encompasses a private responsibility of landowners to prevent damage (German Bundestag 2005: 12), proprietors have to pay adequate attention to the flood risk (Kotulla 2006: 129–30). This paragraph emphasises that flood protection is not only a governmental task (Cormann 2008: § 31a marginal 4), but also landowners have to support flood protection as far as possible.

In final consequence, private landowners of the floodplains individually have to cope with the risk of settling in these locations. A subjective right for protection against financial losses through floods cannot be derived from the law. Landowners are entitled to use their land and others have to let them exercise their rights (Needham 2006: 48). Should the society let landowners exercise their property right clumsily, or are water management agencies obliged to restrict landowners in order to avoid damage?

Water Management Agencies

Water management is the purposeful arrangement of all anthropogenic activities that affect surface or subsurface waters (Knopp in Sieder et al. 2007: preface, marginal 5). Water managers should manage the water system in a sustainable manner. The legal basis for this task is the Federal Water Act (Knopp in Sieder et al. 2007: preface, marginal 2), whereas the administrative competence for water management is a task of the states (*Länder*) (Steenhoff 2003: 51), still after the reform of federalism. What are the duties and responsibilities of water management agencies according to flood protection? The answer to this question helps to understand why levees are still such an important instrument for water management, and why space for the rivers is implemented so inertly.

§ 6 WHG 2010 contains the general principle for water management: water should be operated so that it serves the common good and simultaneously provides benefit for individuals (§ 6 WHG 2010). Since 2005, § 31a WHG has been concretising this principle according to flood protection: damages due to floods should be prevented, water masses should be retained, and harmless discharge of flood waves should be provided. This clause introduces for the first time general and nationwide principles for flood protection (Cormann 2008: § 31a WHG marginal 1, German Bundestag 2005: 12, Schwendner in Sieder et al. 2007: § 28 WHG marginal 2a), whereas the term 'harmless discharge' (*schadloser Abfluß*) refers to environmental and economic goods. The regulation has been introduced in the reformed water law as § 6 WHG 2010. These principles are in particular addressed to water management agencies (Schwendner in Sieder et al. 2007: § 28 WHG marginal 8–9). Regulations according to floodplains and rules for technical river constructions, in particular levees, are relevant to implement these principles (Czychowski 1998: 1172).

Formal designation of inundation zones   Before the Federal Water Act in 2005, the WHG contained no section titled as flood protection. Although, § 31 and § 32 WHG 1996 did mention flood protection explicitly. § 31 WHG 1996 referred to levees, § 32 WHG 1996 referred to floodplains. Preventing flood damage requires regulations for the use of floodplains because water management agencies need to prevent the emergence of damage before a flood event. The relevant regulations have been strengthened in the course of time.

The Federal Flood Control Act in 2005 distinguished two types of submergible areas: first, the § 31b I WHG 2005 defined floodplains as areas between levees or within the river valley or further areas, which can be inundated or overflowed due to floods. These are floodplains in the sense of the Federal Water Act. In addition, areas for retention are also floodplains in this sense. The new Federal Water Act contains the same definition in § 76 WHG 2010. The second type of floodable areas was determined in § 31c I sentence 1 WHG 2005: *überschwemmungsgefährdete Gebiete* – potentially floodable areas – have been defined as areas which can be inundated by a flood, but which need not be formally stated. Such areas encompassed also areas in the river valley, which could be inundated due to a breakdown of flood protection (e.g. a crevasse). The category 'potentially floodable areas' has been criticised as a weak category, unable to provide space for the rivers or prevent damage (Reinhardt 2008: 356, Rolfsen 2009: 770). But instead of strengthening the category, it has been abandoned in the WHG 2010. As the 2005 reform missed the opportunity effectively to confine building in formally determined floodplains (Kotulla 2006), the WHG 2010 missed regulations for potentially floodable areas.

But formally designated floodplains in the sense of § 76 WHG 2010 have been strengthened in the course of time: up to 2005, § 32 WHG 2005 determined that floodplains should be formally designated if the overall situation requires such a designation (§ 32 WHG 1996). Regulations had to be made at the level of the federal states; according to the details of such regulations, the federal states could decide on their own (Gieseke et al. 1992: 1024). Although the reform of the Federal Water Act in 1996 introduced additional clauses according to formally declared floodplains, the legislator in 2005 still relied on the competence of the federal states. This finally led to an inhomogeneous implementation in the federal states (UBA 2003: 150). In 2002, big inundations along the River Elbe produced enormous damage. After the water discharged and relief measures were accomplished, flood protection in Germany was evaluated. Deficits in regulation and implementation were identified (German Bundestag 2005: 1, Stüer 2004: 415).

In order to solve these deficits, the Federal Government reformed the Federal Water Act in 2005. The reform of the article about floodplains contained improvements of the regulatory opportunities as well as enforcements for implementation. § 31b II sentences 3–5 WHG 2005 formulated a period for water management agencies to determine floodplains legally. The period should end in 2012, and in 2010 for highly sensitive areas (Cormann 2008: § 31b WHG marginal 18). For these legally stated floodplains, special restrictions for land use had to be made by the federal states. These restrictions ought to contribute to ecological quality of the floodplains, anticipate erosion, support restoration of floodplains, consider aspects of flood discharge, reduce damage and regulate sewage disposal, water supply and licences for digging in floodplains (§ 31b II sentences 6–7 WHG 2005). The federal states needed not declare every floodplain, which matches the definition in § 31b I WHG 2005; rather it remains to the federal states to declare rivers and parts of rivers for which formal floodplains must be stated (Paul and Pfeil 2006: 506). These rivers and parts of rivers can encompass unsettled areas and settled areas (Cormann 2008: § 31b WHG marginal 10). Additionally, building permissions in the formally stated floodplains were subject to an additional approval by water management agencies (Cormann 2008: § 31b WHG marginal 32). Furthermore, local self-government was restricted in the legally stated floodplains by the WHG 2005 (Breuer 2006: 620, Cormann 2008: § 31b WHG marginal 31 and 34, Kotulla 2006: 131): municipalities ought not to designate these areas for building purposes any longer (§ 31b IV WHG, Cormann 2008: § 31b WHG marginal 30). The restrictions for these areas encompassed even not yet legally stated floodplains, if they were already depicted in special maps (§ 31b V WHG 2005). So finally, in stated floodplains, or in floodplains which are planned for such legal designation, flood protection had priority over other issues (Rolf-Peter Löhr in Battis et al. 2007: 171). Potentially floodable areas ought to be declared and mapped (§ 31c I sentence 2 WHG 2005). The declaration of such areas ought to make people and agencies aware of the flood risk (Paul and Pfeil 2006: 506). According to the legal consequences, the regulations for potentially floodable areas were at the same level as the regulations for formally stated floodplains before 1996. An intention for the Federal Flood Control Act was to provide space for the rivers (German Bundestag 2005: 8). Therefore, floodplains should be restored to retain water (Cormann 2008: § 31b WHG marginal 18). Flood Control Plans (*Hochwasserschutzpläne*) had to be prepared at the level of the federal states to identify and declare factual and potential retention areas (§ 31d WHG). The plans should aim at restoring retention areas (Cormann 2008: § 31d marginal 2). Whether the regulations after the Federal Flood Control Act indeed lead to more space for the rivers has been seen as a 'crucial point' of adoption of the law at the level of the federal states (Moss and Monstadt 2008: 73).

It could be doubted whether the Federal Flood Control Act indeed provided effective instruments to make space for the rivers (Reinhardt 2008, Rolfsen 2009). It was a critique addressed to the former legislation that granting legally binding status for floodplains against regional and municipal land-use planning was a very time-consuming and complex administrative procedure. The new law has neither responded nor solved the problem with the formal designation of inundation zones (Moss and Monstadt 2008: 71–2). For that reason, it was doubted that the inundation zones would have been formally designated within the prescribed period, particularly because delays caused no legal consequences (Kotulla 2006: 130). The WHG 2010 sustains this critique, since a new deadline is set by the legislator (until 2013) (§ 76 WHG 2010), which again missed any penalty for not meeting the deadline. A rule that was abandoned in 2010 was that floodplains must only be declared formally for rivers or parts of rivers for which high damage is expected or has already occurred (§ 31b I WHG 2005). This rule could have led to an absurd situation: areas that are appropriate to give them to rivers, because they are in rural areas with low settlement density, have not to be declared formally. Thus, these areas are legally appropriate for additional settlement development. An application in this manner would cause urban sprawl along rivers and hence would reduce available space for the rivers. Since 2010 whole catchment areas have to be regarded for the formal designation of inundation zones (§ 74 and 75 WHG 2010).

The reasons for the Federal Flood Control Act 2005, as also for the reform of the Federal Water Act in 2010, have been deficits in regulation and implementation in the water law, as the federal government announced. Have such deficits been alleviated through the reforms? The improved regulations support water management agencies in preventing the emergence of flood damages by restricting the accumulation of values in floodplains respectively prescribe adaptive measures. In addition, the regulations for floodplains support the claim for harmless discharge for flood waves, because the emergence of new obstacles for discharge can be avoided. However, the regulations affect only the emergence of new potential damage and other structures in the floodplain, but they do not affect existing values, which can be destroyed. In addition, it is not yet prohibited at all to accumulate immobile values behind levees (Kotulla 2006: 135), so it can be expected that the vulnerability of the floodplains – particularly behind the levees – will still increase. The reform in 2010 failed in improving regulations for potentially floodable areas, whereas floodplain restrictions seem to be at least viable to confine additional settlements in these areas. Building in floodplains meets municipal interests and the implementation of the regulations and decisions about exceptions rely on the responsibility of municipal water management agencies, which is a difficult situation for local decision-making (Luhmann 2005: 24). New urban developments are much more difficult to realise facing the new regulation, and the preservation of open space in floodplains for purposes of retention has essentially been improved. However, no instruments have been provided either to reduce existing potential damages or to provide additional space for the rivers (which is currently cut off from the river by levees or other technical constructions).

Technical river construction (Gewässerausbau) Besides the designation of a legal declaration for floodplains, a further task of the water management agencies is to plan, build and manage constructions like reservoirs, levees or polders, which is a general task of water management (Cormann 2008: § 31b marginal 19, Czychowski 1998: 1171, § 31 WHG marginal 8). These measures are technical river constructions (*Gewässerausbau*) in the sense of § 67 WHG 2010, formerly § 31 WHG 2005. This includes building, removal or reshaping of rivers or levees of rivers (Spieth 2008: § 31 WHG marginal 34).

These measures require complex administrative procedures. In general, a formal plan approval (*Planfeststellung*) is required for all technical river construction measures (§ 68 I WHG 2010). The specific procedures of such plan approvals are determined in the respective state laws (Spieth 2008: § 31 WHG marginal 58). Any change of a levee that affects the flood discharge requires such a formal plan approval, no matter whether a levee will be built, removed or reshaped, and whether it protects a particular area, reduces or accelerates the discharge (Spieth 2008: § 31 WHG marginal 42). In addition, an Environmental Impact Assessment (EIA) is mandatory for levees (Spieth 2008: § 31 WHG marginal 40). The EIA is determined in a particular law (*Gesetz über die Umweltverträglichkeitsprüfung*, UVPG). In addition to a formal plan approval and an EIA, supralocal measures in most states (except the federal states Bremen, Hamburg and Berlin) require a special regional planning procedure (*Raumordnungsverfahren*) in accordance with § 15 ROG (Spieth 2008: § 31 WHG marginal 45). Finally, each change of the shape of a river requires extensive administrative procedures, which cost money, time and it requires negotiated give-and-take.

The WHG determines a standard for technical river constructions: rivers in a natural shape and condition should be maintained in this state, others should be restored to an at least approximately natural state and condition (§ 67 I WHG 2010, formerly § 31 WHG 2005). This rule was introduced in the aftermath of the flood catastrophes in 1993 and 1995 (Berendes 2008: § 31 WHG). This precept ought to care for space for the rivers. However, the implementation of this precept is subject to balancing, confined by the principle of proportionality (Spieth 2008: § 31 WHG marginal 5–7). The precept for the restoration of rivers that are not in a natural or approximately natural shape does not imply a duty to refurbish the river to a historic shape, because excessive burdens should be avoided (Spieth 2008: § 31 WHG marginal 14–15). If the natural or approximately natural shape of a river already equates to the designations in a formal approved plan (according to a *Planfeststellung*), this shape has to be maintained by the water management agencies. A change of this shape would require a new formal plan approval (Spieth 2008: § 31 WHG marginal 12). The maintenance of rivers is determined in § 39 WHG 2010, formerly § 28 WHG 2005. It should preserve the 'proper discharge' (*ordnungsgemäßer Abfluß*) of rivers. Maintenance requires no specific plan approval, EIA or *Raumordnungsverfahren*. Contrary to technical river constructions, measures for maintenance can be implemented without excessive bureaucratic procedures. Therefore it is quite essential whether a measure is a technical river construction in the meaning of § 67 WHG 2005 (respectively § 31 WHG 2005) or just maintenance in the meaning of § 39 WHG 2010 (respectively § 28 WHG 2005).

§ 31a I WHG 2005 emphasised the importance of flood protection for water management (Schwendner in Sieder et al. 2007: § 28 WHG marginal 13e), which was confirmed in the 2010 reform of the water law (§ 6 WHG 2010). Thus, maintenance must encompass flood protection. Otherwise, official liability can be claimed by riparian landowners (Schwendner in Sieder et al. 2007: § 28 WHG marginal 13e). So, water management agencies are interested in quick and less bureaucratic implementations of necessary restorations of levees. This, however, is not possible as constructional river management under § 67 WHG 2010 respectively § 31 WHG 2005. For that reason, restoration, renewal and technical improvements of levees are covered by § 39 WHG 2010 respectively § 28 WHG 2005 as river maintenance measures (Schwendner in Sieder et al. 2007: § 28 WHG marginal 13e). 'Proper discharge' is determined by the first erection of the levee (Schwendner in Sieder et al. 2007: § 28 WHG marginal 13e). § 28 WHG 2005 and § 39 WHG 2010 obligate to maintain this state (Schwendner in Sieder et al. 2007: § 28 WHG marginal 36). Raising the level of a levee, fundamental changing of the shape, especially if the land was not, or not so often, inundated before, are technical river constructions according to § 31 WHG 2005 respectively § 76 WHG 2010 (Schwendner in Sieder et al. 2007: § 28 WHG marginal 13e). Finally, measures for maintenance are more likely to be implemented. They are less complex. Complexity, conclude Timothy Moss and Jochen Monstadt (2008: 318 and 333) in their evaluation of the implementation of flood policies in Europe, is the most important obstacle for restoring floodplains.

In conclusion, flood protection became a more important task for water management agencies since the Federal Flood Control Act in 2005. Their responsibility in this matter has been strengthened by the reform in 2010. Legal instruments are available to cope with this task. However, the instruments are useless without the ability and willingness of water management agencies for implementation (UBA 2003: 136). So, the law sets incentives for water management agencies to maintain the existing situation at a river instead of providing new space by reallocating levees.

Policymakers

In this research, the term policymaker stands for all legislative parties at a higher level than the municipal level. In particular, this includes the legislative bodies at the level of the European Union, the Federal Republic of Germany and the federal states. They influence flood protection by legislation. Thus, these legislations will be presented in the following sections. The relevant reforms and revisions will be explained as well.

Legislation for flood protection   The Federal Building Code (BauGB), the Federal Regional Planning Act (ROG) and the Federal Water Act (WHG) are most relevant for flood protection. The years since 1993 are important, because particular revisions of these laws took place since then (Moss and Monstadt

2008: 63). In the following sections, the legislative competences for flood protection will be presented. Afterwards, the most important reforms and revisions will be discussed.

Who is empowered to legislate for flood protection? The reform of federalism in 2006 changed the legislative competence for regional planning and water management. Before 2006, water management and regional planning was assigned to federal framework legislation (*Rahmengesetzgebung*) (Knopp in Sieder et al. 2007: 3d). However, since the reform of federalism abandoned federal framework legislation, these realms are now an issue of concurrent legislation (article 74 GG no. 31). Concurrent legislation entitles the Federation to rule an aspect or rely on the federal states to legislate (Seiler in Epping and Hillgruber 2008: Art. 70 GG, marginal 20). In principle, the reform of federalism gave legislative competence in the field of flood protection and water management as well as for regional planning to the Federation (Moss and Monstadt 2008: 68), but the federal states may create far-reaching exceptions (article 72 III GG, Moss and Monstadt 2008: 73). This rule is a result of the reform of federalism in Germany (Seiler in Epping and Hillgruber 2008: Art. 70 GG, marginal 23). It should provide flexibility for the adaptation of federal law in the federal state (Seiler in Epping and Hillgruber 2008: Art. 70 GG, marginal 24). However, a less transparent legal system can be expected (Seiler in Epping and Hillgruber 2008: Art. 70 GG, marginal 29). Still, after the reform of federalism, the implementation of federal legislation depends on the implementation in the federal states. Timothy Moss and Jochen Monstadt conclude that:

> owing to strong federalism, national activities were rarely launched and regulations in land-use planning and flood-risk management remained different from state to state (2008: 64).

In consequence, integrated river basin approaches were very rare in the past, and actually very few schemes for restoring floodplains were put into practice (Moss and Monstadt 2008: 64). The legislative competence for flood protection, thus, remains in a concurrence between the federation and the federal states, whereas the federal states have far-reaching power.

Reforms and revisions   As already mentioned, reforms and revisions for improving flood protection usually take place after flood events, not only in Germany but also in other countries (e.g. Johnson and Priest 2008: 513). The most important German legislation in this context is the reform of the Federal Water Act in 1996 and the Federal Flood Control Act in 2005. The WHG revision in 1996 was a result of the 1993 flood at the Rivers Rhine and Mosel (German Bundestag 1996). The Federal Flood Control Act is a legislative reaction to the floods in 2002 in the catchment area of the River Elbe and the River Danube (German Bundestag 2005). In addition, the directive 2007/60/EC of the European Parliament and of the Council of 23 October 2007 on the assessment and management of flood risks must be considered as relevant legislation. The most recent reform of German water law resulted in a completely revised and rearranged WHG in 2010. It is actually a result of a failed attempt for a comprehensive environmental law (*Umweltgesetzbuch*). Due to the reform in 2010, the German government wanted to strengthen water management (German Bundestag 2009: 1). Water law was criticised in the past since it was fragmented (Schneider 2005), decentralised and complex (Moss and Monstadt 2008). The German federal government strives at a more centralised and consistent legislation on water issues (German Bundestag 2009: 1). The reform of federalism in 2006 made this attempt possible (German Bundestag 2009). The section about flood protection has fundamentally been reformed, which was predominantly a reaction to translate the European Flood Protection Directive 2007/60/EG into German law. All reforms and revisions were intended to relieve flood protection of its deficits of implementation (German Bundestag 1996: 1, German Bundestag 2005: 1, German Bundestag 2009: 1), and regulation (German Bundestag 2005: 1). So, the legislations aim at an effective flood protection (European Union 2007: 27, German Bundestag 1996: 1, German Bundestag 2005: 1).

Legislators pursue two major issues due to reforms and revisions of water law: first, flood risks ought to be managed catchment-wide, and in order to make space for the rivers the legislators agree since 1996 on the necessity to restrict the land use in the floodplains. In 1998, the conference of ministers responsible for regional planning (MKRO) claimed that regional planning should use their planning instruments to restore retention volume along the rivers and that urban development should be constrained in flood-prone areas (MKRO 1998: 1). Legislators are thus in accordance with the academic consensus (Greiving 2002, 2003, Grünewald 2005: 14, Karl and Pohl 2003: 267–70, Patt 2001: 57, Petrow et al. 2006: 717, Voigt 2005). How does legislation reflect these ideas?

Catchment-wide cooperation was already demanded in 1995 by the LAWA (inter-ministerial conference of ministers, responsible for water issues in Germany), which votes for a catchment-wide implementation of the principle 'upstream protects downstream' (LAWA 1995: 1). So, the revision of the Federal Water Act in 1996 aimed at consistent land use restrictions in the floodplains (Czychowski 1998: 1170). Also, the federal government of Germany claimed for the area-wide formal declaration of floodplains, including appropriate regulations to reduce flood risk. The implementation of this claim is an explicit aim of the Federal Flood Control Act (German Bundestag 2005). In addition, the EU founded their directive with the argument that coordination between member states was inevitable for flood risk management of the future (European Union 2007: 27). Floods have transboundary effects. This advises the legislators to regulate flood protection to a quite far-reaching extent at a high legislative level (Reinhardt 2008: 469). However, as presented earlier, the factual implementation of the Federal Flood Control Act was a crucial point of the application in state law (Moss and Monstadt 2008: 73). The federal states had far-reaching competences, which allow them to differ from the federal legislation. This weakened the Federal Flood

Control Act particularly, because it has been doubted that the regulations will consistently be applied in practice. The effect of the EU directive is also crucial, because the concrete and measurable targets were not defined in the necessary depth (Reinhardt 2008: 469). Michael Reinhardt speaks in this context of the burden to concretisation, which relies on the member states (2008: 469). Finally, it can be concluded that the legislative parties ought to rule flood protection at a high legislative level, but in fact, the concretisation of the legislation still relies on the subsequent administrative levels, even after the reform in 2010. To some extent, this discrepancy seems to be typical legislation for flood protection: political and academic parties claim concrete and far-reaching legislation; general regulations, which must be concretised by subsequent legislators, are the outcome. This is remarkable because in other realms of legislation, legislators prove their ability to regulate quite far at a relatively high level. The European water quality directive (2000/60/EC) is a prominent example of this (Reinhardt 2008: 469).

The flood-relevant reforms and revisions aim also on the idea that flood management requires restrictions for the use of floodplains. Therefore, § 31b WHG 2005 extended the possible regulations for floodplains (German Bundestag 2005: 9). § 78 WHG 2010 specifies and determines these regulations more in depth. Indeed, restrictions affect the land use. Planning institutions at municipal level are responsible for binding designations of land use (§ 1 BauGB). Nonetheless, § 31b WHG 2005 as well as § 78 WHG are assigned to the responsibility of water management agencies. Rüdiger Breuer criticises the fact that the Federal Flood Control Act had intervened into water rights, but not essentially into building rights. In this context, he names the Federal Flood Control Act as '*Schlag ins Wasser*' [literally: a hit into the water] (Breuer 2006: 617–18). In addition, water management agencies are obliged to declare floodplains formally with restrictions for land use within a legally determined period, but for non-performance of this duty there are no consequences determined (Kotulla 2006: 130). Moreover, in 2005, either this duty has not been applied for such areas, which could be or were actually harmful affected by a flood. Thus, the water management agencies, even if they are obeying the determined deadline, are not obliged to designate formal inundation zones in rural areas (Heemeyer 2007: 78). Nonetheless, rural areas are most appropriate to provide space for the rivers by restricting the land use there. The legislator has not repaired the lack of implementation in flood protection in 2005. But, as mentioned, this regulation has been abandoned in 2010. Unfortunately, the potentially floodable areas have also been abandoned, although this instrument could have been strengthened and used to reduce value accumulation behind levees.

In 1996, the same pattern as in 2005 was observable. Although the German parliament wanted to intervene in spatial planning decisions (German Bundestag 1995: 6), the Federal Water Act was reformed instead of planning law. Also in 1998, after the Oder flood in 1997, the German Bundestag debated about river construction and flood protection. The party *Die Grünen* claimed that interventions in the land use of floodplains by regional planning and land use planning would be necessary (German Bundestag 1998: 1–2). The federal government answered that the 1998 reform of the Federal Building Code and the Federal Regional Planning Act enables spatial planning to designate certain areas with a priority or precedence areas for flood protection. Nonetheless, this designation relies on the competence of the particular land use planners. The federal government does not substantially intervene in their decision-making process (German Bundestag 1998: 19).

The EU directive does not explicitly mention detailed targets according to restrictions of land use in the floodplains. With respect to the principles of subsidiary and proportionality, the European Commission argues that a:

> flexibility should be left to the local and regional levels, in particular as regards organisation and responsibility of authorities (European Union 2007: 29).

The burden of concretisation, as Michael Reinhardt denotes it, relies on the executive and judicative (2008: 470). Michael Reinhardt assumes that the European politicians wanted to avoid debates about details of the directive, but rather politicise the flood issue for election campaigns (Reinhardt 2008: 469).

Has the legislation failed to achieve a more effective flood protection? The intention of the reforms and revisions were to achieve a catchment-wide flood risk management and obliging restrictions for the land use of the floodplains. Deficits in regulation and implementation ought to be alleviated. The legislation seems to succeed to a certain extent, to some extent, however, it can be doubted that the deficits are indeed be solved. Homogeneous catchment-wide flood management and restrictions in land use to make space for the rivers are the desires of policymakers. But owing to an entrenched belief in subsidiary, legislators failed to implement strong hierarchical rules to enforce their desires. Rather, general and non-enforceable rules have been made at the EU and federal level. Instead of intervening in binding land use planning, interventions have been made in water law. Should politicians be voted out, because of their weak legislation? Are they to be blamed? Article 30 GG (Basic Law) provides state power and the discharge of state functions as a matter of the federal states, Article 28 GG guarantees local self-government. These articles sustain strong positions for the respective lower level: as far as each state acts within the law, federal states have a good change to oppose the federation, municipalities have a good chance to stand up to the federal states. For example, the Federal Flood Control Act is a compromise between the federal parliament (*Bundestag*) and the federal council (*Bundesrat*) (Kotulla 2007: 38). Contentious aspects were the strong restrictions for agricultural land use and the discussed prohibition for building in floodplains (Reinhardt 2008: 469). Half a year after enacting the law, the compromise was weakened due to the reform of federalism. The new competences after 2006 produce a contention between the federal states and the German federation (Kotulla 2007: 42). A similar contention exists between the European Union and its member states, since it is a confederation of sovereign member states. Each has to implement directives into state law. So, EU policymakers should not be criticised for too diffident legislation. Moreover, the current legislation is a result of a system-inherent mechanism of contention between the policymakers. Policymakers at a higher legislative level must rely to some extent on the ability and competences of the respective lower level that the legislation will be implemented in a meaningful way.

Why did legislators not change the planning law but water right in 1996, 2005 and 2010? Eventually, the target was in each reform to make space for the rivers, which is an issue of spatial planning. What distinguishes water right and planning law? Whereas spatial planning is metadisciplinary and superordinary, water right is sectoral planning (Stüer 2004: 416). Flood protection is an ordinary and central task of water management (Janssen 2005: 451, Stüer 2004: 416). Spatial planning balances different spatial requirements, emerging from different sectoral rights (§ 1 VII BauGB and § 1 I ROG). Spatial planning has to decide on spatial interests on an objective basis, ideally without forming its own spatial interest (Krautzberger 2007: marginal 28). Thus, the legislator must influence spatial planning decisions by intervening in sectoral planning law. An advantage of this procedure is that the regulations in the sectoral law can bind other sectoral planning institutions (Stüer 2004: 416). Nonetheless, the contenting relation between sectoral planning and spatial planning enables local authorities to balance flood protection with other issues and decide against flood protection (Stüer 2005: 2991). Besides this quite legal-oriented argumentation, there is another reason why policymakers are not likely to intervene in municipal planning processes. Policymakers reason the reforms and revisions of water law with deficits in regulation and implementation (German Bundestag 1996, 2005 and 2010). A more or less anonymous institution is blamed, namely the water management agencies. If policymakers would blame spatial planners for taking space from the rivers and providing it for urban development, they would implicitly blame the victims of a flood for settling in floodplains. This would not be good for an election campaign. However, this is possibly only a side effect.

Flood protection is not only dependent on the vertical implementation due to all legislative and decisive levels; it also depends on the professional balancing process. Reforms and revisions are only possible within a well-established system of balances. Flood protection requires intervention in sectoral law, and relies on wise balancing in practice.

Land Use Planners

§ 1 of the Federal Building Code determines the role of urban land use planning (*Bauleitplanung*). The function of urban land use planning is the preparation and regulation of parcels of land for building purposes (§ 1 I BauGB). Two questions are of interest according to land use planning at the municipal level. To what extent can local planning authorities decide what their course on flood protection will be, and to what extent are they bound in their decisions? Which legally founded incentives drive municipal planners to designate floodplains for building purposes?

Planning in the floodplains   Although the 2010 reform of water law restricts building in inundation zones, particularly zoning the levee-protected floodplains for urban development is not in general prohibited – the levees pretend security against inundations. New building in such areas, however, require special consideration of the flood issue (Stüer 2005: marginal 3454). Which particular considerations do planners have to take into account, under which conditions are planners able to permit building in floodplains?

Preparatory and binding land use plans (*Flächennutzungsplan* and *Bebauungsplan*) are important instruments to prepare and steer urban development. Land use plans have to consider certain issues in their planning to a certain depth. In addition, land use plans have to be adapted to the priority objectives of regional planning (§ 1 IV BauGB). The Federal Building Code determines not only procedural aspects but also contents of planning:

> The planner has to balance and combine the alternatives in order to achieve the best result (Davy 1997: 247).

Land use plans should achieve a sustainable urban development, including social, economic and environmental aspects as well as intergenerational justice. The land use should be socially fair, and the plans should care for a human environment and protect the natural resources. Protection of the climate and urban design should be considered as well (§ 1 V BauGB). All public and private interests must be balanced fairly among and against each other (§ 1 VII BauGB). If two or more issues are incompatibly conflicting with each other, planners must 'balance issues away' (*Belange wegwägen*) or shift it to other administrative procedures, such as, for example, determining special requirements for building permits (Paul and Pfeil 2006: 510, Stüer 2005: marginal 3454). In principle, municipalities are quite free in their decision, to balance one aspect with more weight than other aspects, as long as it is not obviously disproportional (Krautzberger 2007: preface to § 1–13a BauGB, marginal 27–8, Paul and Pfeil 2006: 509).

Land use plans must pay specific attention to a long list of issues, including – since 2005 – flood protection (§ 1 VI no. 12 BauGB). However, it is very likely that flood protection contradicts other issues, because it restricts the economic land use (Stüer 2005: marginal 3450). Flood protection is closely related to effects on the environment, in particular on water in the meaning of § 1 VI nr. 7a BauGB. For these kinds of issues, a formal environmental impact assessment is required; the municipality may determine the depth and level of detail of this assessment (§ 2 IV BauGB). Municipal planners have a certain scope of considering flood protection in land use plans (Paul and Pfeil 2006: 508). Whether an issue is more significant than flood protection, cannot in general be derived from law. Each case must be regarded individually (Paul and Pfeil 2006: 509). Land use plans must insofar consider and balance many different and sometimes conflicting issues.

In addition to balancing issues, municipalities are obliged to coordinate their plans with adjacent municipalities (§ 2 II BauGB). This duty is confined to urban planning aspects. According to flood protection, this could mean that downstream municipalities are forced to build additional constructional measures for flood protection, which affect the urban issues of this municipality. An insignificant aggravation of flood risk is not an affection of other municipalities in the sense of § 2 II BauGB (Paul and Pfeil 2006: 511). Nonetheless, disadvantageous effects of particular planning decisions on third parties, in particular other municipalities down- or upstream, might cause the abrogation of this decision (Stüer 2005: marginal 3460). In addition, inappropriate consideration of flood issues in land use plans might cause official liability (Schneider 2005: 179). Whereas not each incorrect planning decision leads automatically to an official liability, moreover, the fault must be unambiguous (Stüer 2005: marginal 3460).

In some cases, sectoral planning precedes municipal spatial planning. According to § 31 II WHG, in particular sentence 4, formal designations of inundation zones may include settled areas. In addition, it is interdicted for municipalities to develop new binding land use plans in such designated inundation zones (§ 31b 4 WHG, Cormann 2008: § 31b WHG marginal 30). Indeed, this confines the municipal planning authority, but the federal constitutional court has judged that this constraint is not disproportional, because municipalities have to consider natural circumstances – like high water – in their plans (Paul and Pfeil 2006: 506). Finally, due to the Flood Control Act, flood protection is prior to building (Löhr in Battis et al. 2007: 171, Stüer 2005: marginal 3459).

Besides land use planning, a further possibility for municipalities is to permit building in accordance with § 34 BauGB. Building permits can be granted in formally determined urban areas (*Innenbereich*) (§ 34 BauGB). In formally designated inundation zones, these building permits require additional permissions by the respective water management agency (§ 31b IV sentence 3 WHG 2005). The possibilities in these areas for municipal planners are quite confined by the sectoral planning right.

Superficially, regional planning is able to steer municipal planning (Stüer 2005: 2986). Formal regional planning aims (*Ziele der Raumordnung*) are prior in municipal planning (§ 1 IV BauGB). Principles of regional planning (*Grundsätze der Raumordnung*) are subject of balancing. They are not in general prior with respect to other issues (Schneider 2005: 169). The Federal Flood Control Act of 2005 introduced regulations in the ROG, which ought to produce more recognition for flood risk. Especially if floodplains are declared as priority issues of planning (*Vorranggebiete*) (§ 7 IV no. 1 ROG), or as issues of precedence (*Vorbehaltsgebiete*) (§ 7 IV no. 2 ROG), these declarations affect the municipal planning (Paul and Pfeil 2006: 507). But municipalities try to defend the regional planning designation in particular areas. For example, the municipality Heinsberg near Aachen wants six particular areas not to be designated for precautionary flood protection in the regional plan. They were afraid of being confined in their independent planning decisions (*Aachener Zeitung* 2008). Indeed, regional planning must not regulate too far. The more detailed, and the more binding the objectives are, the more crucial is the legitimation (Stüer 2005: 2986). Regional planning must provide some scope for sectoral and municipal planning for concretisation of the contents of regional plans (Stüer 2005: 2988). In cases of conflict, regional planning, as well as sectoral planning, may insist on their supralocal respectively sectoral competences (Stüer 2005: 2990), but even then, municipal aspects must be considered in the regional respective sectoral planning decisions (Stüer 2005: 2995). Finally, nobody is entitled to decide exclusively, but everyone has some say (Stüer 2005: 2991).

In final consequence, zoning floodplains for building purposes is possible. In particular, areas behind levees are still suitable building plots according to the law. Sectoral and regional specifications are not insuperable, but sectoral and regional planning both are able to confine municipal planning to some extent. Even when regional planning imposes restrictions, their implementation in local land use planning needs time. A study about dealing with natural hazards in German spatial planning concluded:

> urban land use planning is up to now normally not restricted or just influenced
> by the new regional objectives (Greiving 2006: 73).

Every planning decision underlies the precept of balancing, no matter whether municipal, regional or sectoral planning authorities dominate in a particular case. Planning decisions and balancing stick together like two sides of a coin. Article 14 GG (guarantee of private property), article 28 GG (local self-government) and article 20 GG (rule of law and precept of democracy) are the constitutional basis of this relationship (Stüer 2005: 2995). After all, 'only a limited number of state regional planning acts include planning provisions for securing new floodplains'. (Moss and Monstadt 2008: 66). Municipalities still designate floodplains for building purposes (Moss and Monstadt 2008: 75).

What are then the motives to designate floodplains for building purposes? Mayors and municipal planners could have an easy job if they would just refer to sectoral right and regional planning, which restrict building in floodplains and keep floodplains undeveloped.

Municipal motivations   'Land for building is scarce' Egbert Dransfeld and Frank Osterhage begin their expertise about the correlation of population and municipal budget (2003: preface). They conclude that municipal budget depends to a high degree on the number and structure of population. Thus, they recommend municipalities to mobilise building land for young families (Dransfeld and Osterhage 2003: 54).

Municipal budget consists of three groups: taxes, fees and financial allocations (Dransfeld and Osterhage 2003: 16). Fees can be neglected for the question pursued here. The financial participation of municipalities in taxes is determined in article 106 GG. Hence, municipalities receive a fraction of income tax and turnover tax (article 106 V and Va GG). Land taxes, trade tax and excise taxes are fully assigned to municipalities (article 106 VI GG). In sum, one-third of municipal incomes are taxes; the most important taxes are the trade taxes and the income tax

(Dransfeld and Osterhage 2003: 16). Municipalities may regulate the collection rate (*Hebesatz*) of the land taxes and trade tax (article 28 II sentence 3 GG). The participation in income tax is calculated based on the income of inhabitants in a municipality (Dransfeld and Osterhage 2003: 23, Kube in Epping and Hillgruber 2008: article 106 GG, marginal 33). Financial allocations by the federation and the respective federal states are a further third of the municipal income (Dransfeld and Osterhage 2003: 16). The municipal income is based on the amount of population (Kube in Epping and Hillgruber 2008: article 107 GG, marginal 23).

These are incentives for municipalities to pull young, well-situated families with relatively high incomes as well as growing enterprises (Dransfeld and Osterhage 2003: 54–5). Municipalities need to designate land for building for these target groups. As presented at the beginning, the nearness to the river is an advantage of location, whether for the nice scenic views for the families, or for using the rivers economically. According to the original question, additional usage of floodplains for building purposes is in the interest of communities. Hence, the emergence of new developments is likely; flood protection, however, often contradicts these municipal interests (Schneider 2005: 195). Finally, zoning floodplains for building purposes is not only possible but also likely.

In addition, the BauGB is building oriented (Davy 2006: 27). It emerged in the 1960s, when housing was required. Each instrument in the building code has an inherent message, which promotes additional settlement development. Instruments for shrinking are not available in the German Building Code. This is similar in the Netherlands, as Barrie Needham points out when he describes the Dutch as a 'nation of builders' (2007a: 25–7). The system of planning, law and property rights strives for building and growth.

Legitimated Patterns of Action

The previous four sections showed that the German system of planning, law and property rights legitimates each stakeholder's activities in floodplains. The described iterative patterns of human activity in floodplains are not only in accordance with the law, but the legal system supports these patterns in the profitable, the dangerous, the controllable and the inconspicuous floodplains. From a legal perspective, none of these phases can be said to be illegal.

Profitable floodplains are managed in an individualistic way. The land market frames patterns of action. In particular, land use planners and landowners are active stakeholders. Zoning provides lucrative allocations; investors and developers demand these sites. The invisible hand of the market, competition, supply and demand – these are the essential ingredients in this phase. The law supports the activities of landowners and planners: landowners are entitled to use their property freely and gather the profits from using it (Article 14 I GG, § 903 BGB). For fiscal reasons, municipalities profit from an increase in population and from well-situated inhabitants. The law sets incentives for both, landowners and municipal land use planners continue building in floodplains.

When a flood occurs, when the profitable floodplain changes to a dangerous floodplain, another management strategy is observable. In case of danger, the community closes ranks. The fear of damage drowns the worldview of individualistic profit. Damage due to a flood, however, affects all members of a community. A threatened community stands close together to defend against a flood. The law supports this pattern of action, because nobody in the floodplain is entitled to get compensation of losses due to a flood. Only in cases of official liability, do members of a community have a chance to sue the water management agencies. Water management agencies, which are legally obliged to maintain flood protection measures, defend the danger of a flood with a lot of effort. Thereby, the flood protection is provided to each member of a community – quite egalitarian.

The controllable floodplain becomes apparent in the aftermath of a disaster. Water management agencies retrieve control over the floodplain. Programmes and plans for reconstruction emerge; money will be made available for flood protection. Law and order, control, hierarchies – these are the key elements of floodplain management in this phase. These patterns of action can be explained in terms of rights. In the aftermath of a flood, water management agencies need no longer provide relief, but rather must prepare the floodplain for the next event. Resources for well-considered planning are available, and necessary: extensions, reallocations and other technical measures of river construction require complex administrative procedures, as it is prescribed in water law, but rebuilding and strengthening levees and maintaining the previous shape of the river systems is with respect to the law quite easy to realise. So, levees are going to persist. In addition, the policymakers are under pressure to react to the disaster, because deficits in implementation and regulation become obvious. Reforms and revisions are the rational response. Which laws are changed depends on the legislative competences. These competences are quite distributed among the EU, the German federation and the federal states, as discussed above. Checks, balances and clear hierarchies are important in this phase.

The fourth phase of floodplains, the inconspicuous phase, yields another pattern of action, which is mirrored in the law. In this phase, the flood is managed, the reconstruction is finished and legislation has been adapted. There is nothing else to do, but to wait for the next flood and to recognise the next failures. In this phase, water management agencies are not the dominant stakeholder. Policymakers solve other problems, like bird or swine flu. Municipal planning agencies have to consider their problems. Each stakeholder is busy with implementation. But, legally framed, nobody intervenes in the work of the other. Municipal self-government, legislation at the level of the federal states – these patterns follow legally founded principles. In other words, intervention is not intended. Quite composed and relaxed, each stakeholder relies on the competences of the others. According to the flood issue, each acts somehow fatalistically.

Finally, the German system of planning, law and property rights reflects on how society manages the land in floodplains in certain phases. The law does not provide a rational answer to the clumsy floodplains, rather are rights in land itself a social creation (Needham 2006: 13). The legal system mirrors contesting ideas, worldviews, rationalities. What does that mean? The law does not determine human behaviour; rather human behaviour determines the law. In this manner, law sustains the patterns of human activity by setting incentives, supporting certain actions, punishing other actions. Finally, law matters for clumsiness, but it determines rather a frame of actions, not a path dependency.

Rationalities of the Social Construction

So far, we can conclude that floodplains are complex socially constructed riparian landscapes. The social construction appears in four particular perceptions of floodplains of predominantly four stakeholders in certain phases. Floodplains are sometimes perceived as profitable, dangerous, controllable or inconspicuous. Each of these perceptions of floodplains is linked to a certain pattern of activity. These patterns of human activity iterate in phases. The system of planning, law and property rights legitimates these patterns of human activity. The law even sustains these patterns, even after several revisions. In sum, the complex social construction of floodplains enables the society to cope with the challenges – and even with extreme floods. The social construction of floodplains is quite viable, although it is not perfect. In the long term, however, flood risk increases due to the social construction of floodplains. This seems not to be wise. So, is the social construction of floodplains rational?

The Theory of Polyrationality

> If you're having to ask who's right (worse still, if you already know who's right)
> you're wrong! (Ellis and Thompson 1997: 209).

Social systems are neither rational nor irrational – they are rather polyrational (Davy 2004: 143). Different rationalities appear together in social systems. Each rationality explains parts of the system and none is able to explain the whole system. Rationalities are filters (or 'condoms', Davy 2008) for perceiving a certain situation. These filters determine actions. The actor perceives them as rational responses to the world and so persists in acting in this particular way (Thompson et al. 1990: 39). Can we identify the rationalities to explain the complexity of the social system in floodplains?

Four rationalities   Cultural Theory provides an answer. Cultural theorists are concerned with the question:

> how individuals confer meaning upon situation, events, objects, relationships
> – in short, their lives (Thompson et al. 1990: xiii).

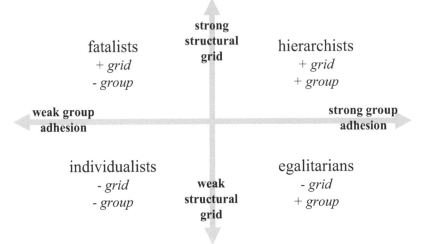

Figure 1.4   The social map

In other words, Cultural Theory is concerned with rationalities (e.g. Schwarz and Thompson 1990: 6–7), whereas ways of life, social solidarities and of course cultures are often used synonymously. Mary Douglas, an anthropologist, founded Cultural Theory; Michael Thompson and others developed it further. Mary Douglas invented a social map, which depicts two dimensions of sociality: Grid and Group (e.g. Thompson et al. 1990). The Group dimension shows an individual's adhesion to groups; it can be weak (left side) or strong (right side); the Grid dimension describes a spectrum of social prescriptions, from 'self-determined' (down) to 'tightly constrained' (up) (e.g. Schwarz and Thompson 1990: 6–7). The social map helps to reduce complex social systems to a simplified model. This model enables us to understand the rationalities, which drive the social system.

Four quadrants emerge. Each quadrant represents a particular rationality, different in their Grid and Group characteristic: the individualistic, the egalitarian, the hierarchic and the fatalistic rationality (see Figure 1.4). The properties of these rationalities are described in various publications, in particular by Michael Thompson, but also by Mary Douglas, Benjamin Davy, Aaron Wildavsky and others (Davy 1997, 2004 and 2008, Douglas 2005, Ellis and Thompson 1997, Schwarz and Thompson 1990, Thompson et al. 1990). The following descriptions of the rationalities summarise some characteristics.

Individualists' rationality: individualists are assigned to the quadrant at the bottom left; weak structural grid accompanied by the conviction that self-determination is an important value. Individualists believe in efficiency, liberty and market mechanisms. 'The strongest will survive' and 'performance pays off' are strong beliefs of this rationality. Individualists are distrusting of the state, neglecting moral claims and they dislike laziness. They emphasise bravery and activity. The invisible hand of the market will bring out the optimal result, as long as nobody intervenes in the free economic forces. Failures happen through unwise interventions.

Egalitarians' rationality: egalitarians believe also in self-determination, but only in strong communities. Moral, social capacity, ethic principles – they are the important issues. The egalitarian leitmotiv is 'all for one, one for all'. Egalitarians establish trusts and civic initiatives. Inclusion and exclusion determines their world. In a market, they act in a group-oriented manner. The competition in the market is between the group and the others – not within the group. The egalitarian rationality neglects rules and the state. Like individualists, egalitarians are afraid of hierarchies, but also of markets. Trust, cooperation and consensus are essential; reckless opportunists, dominant bureaucrats and excluded fatalists help egalitarians to distinguish between the evils and their strong community.

Hierarchists' rationality: hierarchists are located in the social map in the top right quadrant. Hierarchists organise themselves in strong structures with determined rules, and they obey authorities. Therefore, they establish agencies and take up a subordinate role to the system. Law and order, checks and balances are important ideas. Decisions must be well considered. Therefore, expert committees can be useful. The market must be regulated, the egalitarians are obstructers and sceptics, the fatalists must be steered. Hierarchists strongly believe in the controllability of the world.

Fatalists' rationality: fatalists neither align to groups, nor act self-determined, rather they believe in heteronomy; they will not try to intervene at all and accept the situation as uncontrollable: 'it's just fate'. All the others try to rule the world, or believe their moral claims and initiatives can really change, or even rely on the market. It all does not matter. The world just happens – unpredictable. This is the deep belief of fatalists.

Why four? The described four rationalities and their perceptions of the world are still a simplified model of reality. Actually, a fifth rationality, the hermits, is often discussed (Mamadouh 1999: 399–400). Hermits neglect all social interactions and transactions. Their strategic approach to the world is one of withdrawal (Thompson et al. 1990: 29). Therefore, hermits are not relevant for the analysis of human patterns of activity in floodplains. For that reason, they will be neglected here. Cultural Theory does not claim that the arrangement of the four rationalities is the only possible one, but it is a viable approach to simplify complex situations to a manageable number of rationalities (Davy 2004: 145). What makes the theory so useful for the analysis of complex situations is not how many rationalities – four, five, 12, or just two – are involved, but that it neglects blaming irrationality. A theory that argues in terms of rationality and irrationality, as most traditional approaches do (Davy 2004: 144), automatically advocates for the right and the wrong solution of a problem. This, however, hinders analysts to understand how two rationalities interplay, how discourses emerge. Polyrationality, on the contrary, explains disputes, discourses, interactions, negotiations etc. by referring to different rationalities.

Four myths of nature   Each rationality has its own model of ecosystem stability. This explains the observation of ecologists who studied ecosystem management, why different managing institutions:

> faced with exactly the same kinds of situation, adopt strategies based on one
> of four different interpretations of ecosystem stability (Schwarz and Thompson
> 1990: 4).

These models represent different 'myths of nature'; myths of nature explain the diversity of institutional responses to the world (Thompson et al. 1990: 25).

A ball in a landscape usually illustrates myths of nature. The interaction between the ball and the landscape represents the perceived relationship between management and nature. The ball represents 'life', the landscape is the 'world' and the motion of the ball describes 'how life goes on' (Davy 1997: 318). Michel Schwarz and Michael Thompson brought the ecologist's myth of nature in accordance with the anthropologist's rationalities in their Cultural Theory (1990: 6–10). The four myths are the following:

Nature is benign. Humans relate to nature like a ball in a dale:

> No matter what knocks we deliver the ball will always return to the bottom of
> the basin (Schwarz and Thompson 1990: 6).

System inherent forces stabilise the world and lead to a benign equilibrium. Scarcity is only a question of ingenuity. In consequence, individualists believe in the ability to design nature (Zwanikken 2001: 31). Failures of this equilibrium are restrictions for the ball. The ball needs a management based on liberty. This is the deep belief of individualists.

Nature ephemeral is the perception of nature of the egalitarians. Nature is in a state of adaptable equilibrium. Nature is vulnerable. In the picture of the ball and the landscape, the ball lies on the top of a hill and even a small movement would lead to a crash. The egalitarians want to be the protectors of the world (Zwanikken 2001: 31). They act like the principle: prevention is better than recovering (Zwanikken 2001: 31).

The perception of the relation between nature and the humans can be described differently from the hierarchists' point of view: nature is like a wavy landscape and the humans are like a ball in a dale of this landscape. Ecosystem management has to respect the edge of the dale, but it is tolerant as long as the humans stay within this dale (Zwanikken 2001: 31). Wise planning to enforce and rules and institutions to control are needed.

For the fatalists, nature is capricious. For them, nature is more powerful than humans are. Fatalists just use nature without any plan. They do not care about scarcity and they have no idea how to handle resources. They just cope with erratic events (Thompson and Schwarz 1990: 5, Zwanikken 2001: 31). The ball is just in the middle of a plane. It cannot be predicted whether it rolls to the right or to the left.

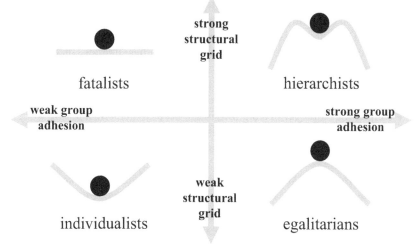

Figure 1.5   The social map and myths of nature

The myth of nature and the rationalities stick together. For that reason, symbols for myths of nature are usually used as symbols for the rationalities. They can be depicted in the social map (see Figure 1.5).

Situation-oriented polyrationality   Can we identify the rationalities to explain the complexity of the social system in floodplains? The social system consists of social interrelations between four groups – landowners, water management agencies, policymakers and land use planners. None of the stakeholders drives the iteration loop in floodplains. As depicted in the analyses of the legal framing, each actor depends on other stakeholders: landowners cannot build without planning permission, water management agencies react on public claims for protection, the interplay between water management and policy drives technical river constructions like levees, the floodplains become inconspicuous due to the reciprocal relation between policymakers and planning. Not the rationalities of the particular stakeholders are of interest, but rather the rationality of the relations; which sounds strange, because a relation of two or more actors is determined by the acting people. However, stakeholders act within an existing social system. This social system influences the relations. This is of interest in this research.

Situation-oriented polyrationality asks how, not by whom rationalities are involved in certain situations (Davy 2004: 144). An actor-oriented approach would require assigning a particular rationality to a particular stakeholder. This will not explain the interplay of two stakeholders in certain situations. Situation-oriented polyrationality regards social systems (Davy 2004: 144). It is a theory that helps to analyse complex situations. Situation-oriented polyrationality avoids what psychologists call the fundamental attribution error. This error is the exaggeration of traits as primary causes for the behaviour of other people, and the underestimation of situation-dependent behaviour (Encyclopedia Britannica 2009, Welzer 2009). To put it incisively, situation-oriented polyrationality assumes that it is not people's actions that determine situations but rather situations that determine people's actions.

That means that the same person – the same stakeholder – can act out one rationality in one particular situation, and another rationality in another situation. For example, the employer in a big firm might act as a hierarchist during work (in relation to his boss), act as an individualist on the motorway on his way home, act as an egalitarian in his family, and as a fatalist when reading the newspaper.

Impossibility theorem   What makes polyrationality a theory? So far, it is only a categorisation of cultural biases, or rationalities (as these terms are used far-reaching synonymously in Cultural Theory and polyrationality). What statement does the theory provide for the interplay of these rationalities? The impossibility theorem provides such a statement: there are just five (including hermits) rationalities, and none of these can ever become uninhabited (Thompson et al. 1990: 86). In other words, social systems sustain the existence of the described rationalities. If one or more rationalities are missing in a social system, it becomes unstable (Davy 2004: 145), because the missing rationalities will emerge and destabilise the existing social system.

This impossibility theorem explains that situations are stable, if all rationalities are present in a 'permanent dynamic imbalance' (Thompson et al. 1990). The rationalities permanently ally with others and simultaneously compete with others (Davy 1997: 319). This permanent dynamic imbalance, states the theory, leads to stability, because polyrational situations are less prone for being surprised by external changes of the settings of a situation. Surprise, however, is the initiator for changing the rationality, and thus for a changed behaviour (Thompson et al. 1990). Nevertheless, if all rationalities sustain certain patterns of behaviour, a change is unlikely. Polyrational situations are less prone to change (Thompson et al. 1990: 96). At least one of the rationalities expected the external change that surprises the others.

This is the reason why polyrationality is helpful to describe the situation in floodplains. The social system in floodplains is robust. Iteratively, the same patterns of human activity persist. We are dealing with a stable social system – is it then polyrational? Are all four rationalities present in the system?

Rational Floodplains

Following the reasoning above, the social system in floodplains needs not to be rational, but polyrational. This makes it more stable. Which rationalities are involved? How do different rationalities establish a robust social construction of floodplains? As mentioned earlier, different perceptions of the floodplain exist:

they are profitable, dangerous, controllable and inconspicuous. These perceptions represent the underlying rationalities.

Individualists' profitable floodplains Building in floodplains is a rational consequence of the perception that floodplains are profitable land. If floodplains are perceived in this manner, buildings must emerge. Therefore, landowners have to be free in their choice of land use. Strong planning regulations hinder the 'invisible hand' of the market, thus planning should support landowners with granting building permissions. Landowners must be able to gather the land rent in order to achieve the best land use. Land use planning is expected to support landowners in this mission (Needham 2007a: 27). The rationality, which sustains this kind of thinking and acting, is the individualistic rationality: less structural grid, less group adhesion. Landowners build in areas at their own risk. The freedom of choice, the 'invisible hand' of the market, will automatically steer to efficient allocations of valuable land uses. Probably, landowners suffer damage, but then it simply was their own wrong choice. Next time will be better (the ball in the landscape will automatically roll in the dale); nature is benign. This kind of 'trial and error' approach (Schwarz and Thompson 1990: 66–7) will work as long as no market failures will be produced due to too strict rules and regulations, and too heavy interventions will mislead in the market. Following this rationality, failures may happen. Floodplain management must therefore dare innovative and risky approaches (such as, for example, floating homes).

This rationality dominates in comparatively dry periods (e.g. in the 1970s and 1980s at the River Rhine, when no extreme flood occurred in this catchment). In this phase, the interplay of land use planners and landowners is able to produce urban land uses in floodplains without serious interference of the other rationalities; floodplains are then used profitably by developers, architects, landowners and municipalities. Along the River Rhine, where the last serious flood events occurred in 1993 and 1995, the profitable floodplains are already observable: the new *Rheingalerie* in Ludwigshafen or the *Rheinauhafen* in Cologne are prominent examples.

Egalitarians' dangerous floodplains Floodplains are dangerous – this statement claims for retraction or protection. If retraction is not possible or very difficult (because the values at risk are immobile), protection remains the only alternative. Protection means to defend against floods. Every landowner in the floodplain is affected by the common threat. Protection is an issue of the community. If the community is not able to protect, it will be very bad for everyone. This thinking matches the egalitarian rationality. Strong communities with a strong group adhesion of members are important. The arguments are neither based on expertise, nor on economy, but on morals. Landowners must be protected with each available resource: sandbags, technical equipment, voluntary relievers, etc. From the egalitarian perspective, protection must be provided also for the poorest. Thus, the community must stand together. Claims for governmental support

(LAWA 2004: 27), demonstrations for levees (Helbig 2003: 129–30) and civil initiatives for flood protection are typical reactions in the aftermath of a flood disaster (like in Cologne-Rodenkirchen or Ludwigshafen-Altrip). The situation of the landowners in floodplains is like the ball on top of the hill. Landowners are at threat of crashing down the hill. Protection is required against this threat. This is the reasoning of water management agencies and landowners for building levees. Mistakes must not occur. Protecting floodplains is a trial without the scope for errors (Schwarz and Thompson 1990: 66–7).

The egalitarian rationality typically results from extreme floods. The individualistic rationality will be drowned by the egalitarian: floodplains are dangerous rather than profitable, victims and voluntary relievers determine the public perception of floodplains.

Hierarchists' controllable floodplains   Water management agencies and policy-makers are then present in the next phase. After cleanup measures, experts analyse the floodplains. Engineering solutions will then typically be worked out to control the flood risk. Policymakers support fast technical solutions. Floodplains are construed as controllable. Water management agencies and policymakers work on flood protection improvements based on regulation, rules and norms. This is a hierarchical approach to floodplains management. Nature is tolerant. As long as a certain threshold will not be exceeded, building regulations are applied, and technical norms are fulfilled, floodplains are controllable. The compliance of all this depends on command and control. Hierarchists have to care for the ball to stay in the dale on top of the hill. Disasters happen, if the ball crashes down. Rationally, levees must be built and improved; building restrictions must be stated.

The hierarchic rationality drowns the egalitarian rationality of the previous phase. Policymakers and water management agencies determine the public perception: floodplains are controllable. In the end of this phase, the parliament passes new laws (like in 1996 and 2005 in Germany, approximately three years after the 1993 flood and the 2002 flood, respectively).

Fatalists' inconspicuous floodplains   Floodplains are inconspicuous. If floodplains are inconspicuous, no special attention must be paid to this land. Floodplains are just land. This perception of floodplains emerges some time after the occurrence of floods, when cleanup has been finished, levees have been reconstructed, heightened and improved, after legislation has passed, when the three active rationalities have done their work. The 'window of opportunity' for flood protection measures is closed again; this window is the period after a flood, when public interest in flood protection is present (Bahlburg 2005: 13). Indeed, legislation is in force to restrict land use in floodplains, but due to subsidiary and local self-government, implementation depends on local and regional land use planning. But floods produce peaks of public awareness. After the peak, implementation is no longer the most important topic on the agenda.

Flood protection becomes one of many topics of spatial planners, which will be implemented or not. This matches the fatalistic rationality.

Other rationalities do not insist on their position in this phase. Stakeholders would denominate this phase as normalisation. As long as policymakers do not enforce local planners to act in a certain pattern, and as long as planners do not claim for better instruments, which are both quite improbable, the fatalistic rationality is present in the floodplains, the hierarchical approach is abandoned. Later on, deficits in implementation and regulation will be asserted (like in the legislative process in 1996 and 2005). Whether floodplains will be used for urban land uses is unpredictable, at least, it depends not primarily on flood protection.

The Polyrational Social Construction

Finally, a polyrational social construction is at work. The bricks of this construction are the patterns of human activity; the perceptions of floodplains are the cement, which represent four rationalities. The construction typically works in phases; the scheme must then be read counter-clockwise (see Figure 1.6).

This social construction pushes increasing risk of flood damage. And it is a self-enforcing system of human activities. The law sustains the social construction. The legal system does not only legitimate the human activities, but also sets restrictions and incentives for the emergence of the rationalities. Who made such 'rules for using land' (Needham 2006)? Obviously, the legal system is not a result of only one rationality. Moreover, with respect to floodplains, law is irrational. It nurses clumsy floodplains. It can be assumed that no legislator intended to sustain clumsiness. The legal system is also a result of a permanent dynamic imbalance

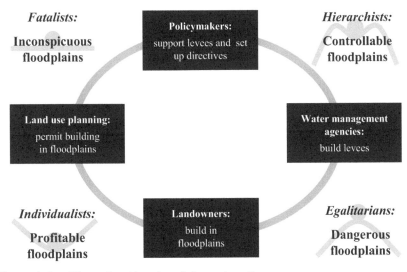

Figure 1.6   The polyrational social construction

between the rationalities. However the law emerges, it finally supports the persistence of the social construction.

The four social actors – landowners, water managers, policymakers and land use planners – commonly manage floodplains with many rationalities instead of one rationality. Rationalities create different storylines of how the world is and how policy should manage it. These storylines and management ideas inevitably contradict each other. They are all based on certain experience and wisdom missed by the others (Thompson 2008b: 25). Uncertainty is a precondition for the emergence of different storylines. The 'social constructions of reality' do not neglect certain 'nursery facts', such as that water does usually not flow uphill (Thompson 2008b: 43). For cases of certainty, different storylines are less influential. But especially in situations of uncertainty, explains Michael Thompson:

> the upholders of each form of solidarity tend to choose those possible states of the world that best support their way of organizing and most discomfort those of their rivals (2008b: 43).

Flood risk, an inherent threat in floodplains, however, contains uncertainty. Rationalities (as abstract social actors) try to persuade other rationalities about how this uncertainty should be treated – by building levees, by relying on fate (which actually no rationality claims loudly, but fatalists pursue this approach), by using floodplains as long as possible, as profitably as possible, or by strong communities, which are able to defend against the water. To promote their storyline, each of the four rationalities raises its voice in the process of floodplain management. Each rationality succeeds in certain phases (whereas the fatalistic rationality does not actively raise its voice, but in the inconspicuous phase of the floodplain, the others are not that dominant). Politics – the planning system, the law and the property rights regime – are responsive to each rationality. This is the conclusion of the empirical study and the analysis of the legal framing. The resulting floodplain management is a process of compromises. Far from ideal, it appears rather clumsy (Verweij et al. 2006a: 8).

> The term clumsy refuses 'the idea that, when we are faced with contradictory definitions of problem and solution, we must choose one and reject the rest' (Verweij et al. 2006a: 19).

This clumsiness 'emerges'; it is not created by design (Verweij et al. 2006a and 2006b, Thompson 2008b). Floodplains provide perfect conditions to observe such clumsiness: uncertainty and responsibility for floodplain management is distributed among many different actors with different rationalities involved. So, the contemporary management of our floodplains is clumsy.

Chapter 2
Coping with Extreme Floods

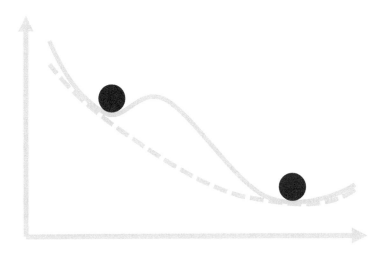

Paradigms for Coping with Extreme Floods

> Reinsurance experts expect annual damage charges from large-scale natural
> disasters to reach between US$ 25 billion and US$ 50 billion (in today's values)
> by the end of the decade (Schwarze and Wagner 2004: 155–6).

We will experience more extreme floods in the future. Are clumsy floodplains
able to cope with additional extreme floods later on? Clumsy floodplains sustain
contemporary flood protection. The policy 'space for the rivers' is difficult to
implement in this context. 'Space for the rivers' is a slogan for the new paradigm:
flood risk management instead of flood protection. Such a flood risk management
will be proposed in this chapter: Large Areas for Temporary Emergency Retention
(LATER). It claims for space for the rivers for extreme floods. A responsive
land policy (Davy 2005: 117–24) is required to change from a levee-based flood
protection to the more efficient floodplain management by LATER. But first, some
basics according to the physics of extreme floods have to be explained.

Extreme Floods

A flood occurs when the water level exceeds a certain threshold. This can be a
flash flood, inundations through torrential rain, surges and river floods (Patt

2001: 2–3). River floods are categorised by probabilities, expressed in annualities – *Jährlichkeiten*: if a flood occurs statistically with a probability of once in a hundred years, it is called a centennial flood (Strobl and Zunic 2006: 383). Water engineers estimate a certain discharge volume – resulting in a certain water level for certain annualities. These calculations are the basis for the size and shape of technical river constructions – particularly of levees. Therefore, a threshold has to be determined.

River floods result from long enduring precipitation, transformed trough retention-capacities of a big catchment area (Patt 2001: 3 and 11). Derived from these definitions, three parameters determine flood events: the properties of catchment areas, the retention capacity and the precipitation (Patt 2001: 11).

Properties of catchment areas   Catchment areas are determined by watersheds. They can be identified based on topographical maps. The catchment area is decisive for flood events. Generally, the size of the catchment area shapes the flood wave. Big catchment areas produce low rising and stretched waves, smaller catchment areas produce fast rising and short-term flood events (Patt 2001: 56). This is because the front of a wave flows faster than the end. So, the distance of flow and the absolute size of catchment areas determine the form and the magnitude of flood waves. The forewarning period of an approximating flood wave depends – besides the technical aspects and the organisation of disaster management – on the size and shape of catchments; it varies between a few minutes in small catchments up to over 12 hours in large catchments (Patt 2001: 8).

Indeed, the size of catchment areas is stable, but humans influence the properties of catchment areas for example through river reconstruction measures. In the past the lower levels of the floodplain along the River Rhine between Basel and Bingen were often flooded (extreme flood events occurred in 674, 886, 1124, 1295, 1342, 1573, 1651, 1652, 1758, 1784, 1813, 1876 and 1882/1883). In 1882/1883, levees broke and 40,000 hectares of the 150,000 hectares floodplain were inundated. Many settlements were destroyed and left by the inhabitants. Malaria, typhus and dysentery spread; navigation on the River Rhine was restricted. In consequence, the landlords decided to reconstruct the river (Böhm and Deneke 1992: 176). In 1818/1919, the engineer Johann Gottfried Tulla (1779–1828) began to reconstruct the River Rhine to make it navigable. The effect was that the average flood levels on the *Oberrhein* (the River Rhine between Basel and Bingen) declined up to 1.5 metres (Böhm and Deneke 1992: 179). The whole area seemed to be better protected against inundations. However, those reconstruction measures had negative effects too. One of the unintended side-effects of the river reconstruction of the River Rhine by Tulla is that the flood waves of the River Rhine now overlap with the flood waves of the River Neckar, the River Mosel and other tributaries, as experienced during the flood events in 1993 and 1995 (Böhm and Deneke 1992: 190). This worsens the flood situation downstream. Finally, intensive uses led to reconstructions of most European rivers in the past two centuries (Strobl and Zunic 2006: 384).

Furthermore, the morphology and the geology are important influences of floods. The soil, the arrangement and form of clefts and valleys, and (to a lower extent) the land use are attributes that influence patterns of flood (Patt 2001: 13–14). But these attributes of catchment areas have changed in the past by measures outside the riverbed. These measures include, for example, consolidation of farmland, irrigation measures, intensification of agricultural land cultivation, or enlargement of the hard surface through extended settlements. In sum, all these measures influence the flood situation on a river. Finally, the whole properties of the surface within a catchment area characterise the wave (Patt 2001).

Retention capacities   Rainfall in a catchment area distributes into evaporation, runoff and retention. Evaporation is negligible in flood situations, so flood events are determined by the relation of runoff and retention. Runoff can be distinguished in surface runoff (rills and runlets), interflow (runoff in upper soil layers) and base flow (deeper and long-distance runoff). Figure 2.1 (Patt 2001: 13) shows how the different runoffs at a given place overlap each other. The base flow stream is invisible under the surface towards the rivers; thus, it serves for a more or less constant level of groundwater. If it rains at this particular place, the rainwater flows to puddles on the surface. Depending on the infiltration capacity, after some time the water begins to seep into the soil. Here, it begins to flow in the same direction as the base flow – except in some specific cases of afflux – the discharge level looks like the dashed line in Figure 2.1. In case of extreme rainfall, the soil cannot absorb all the rain in the subsurface runoff – how much, depends on the infiltration capacity. Thus, the water runs off in rills and runlets at the surface. As is observable after heavy rain on hard surfaces, the discharge rises very fast, and abates after some time (Patt 2001: 13). Flood waves are influenced mainly by the surface runoff and the interflow, the base flow is normally moderately constant.

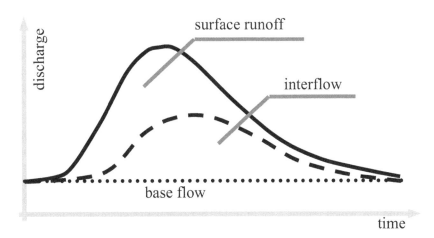

Figure 2.1   **Runoffs at a given location**

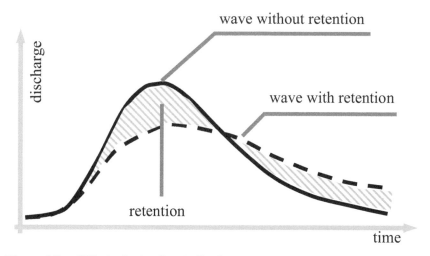

Figure 2.2   Effect of retention to flood waves

In principle, floods are extreme runoffs; retention delays water runoff. Retention is a temporary hold back of water in certain retention areas (e.g. lakes, dams, river basins or floodplains) (Böhm and Deneke 1992: 179). It deforms the flood wave: with the increasing wave, the flood steps over the banks into the retention area, with the discharging wave it flows back into the river. Thus, the peak will be reduced and emerge temporally delayed, as depicted in Figure 2.2. The water masses will not be reduced.

The natural retention capacity in a catchment area results from the four water reservoirs – natural cover, soil, terrain and hydrography (Patt 2001: 18–23). While the natural cover influences extreme floods to a lesser extent, the soil has the highest effect according to its retention capacity. In cases of heavy rain, the actual capacity is finite. Thus, not only a single weather event is decisive, but the weather in the period before as well. Since the winters are usually more humid than the summers, in winter the soil is saturated with water and more water comes to runoff (Patt 2001: 18). In addition, the river itself retains water: due to time of flow, a flood wave deforms. A flood wave consists of a wave rise, a vertex and the wave discharge. The rise is faster than the discharge and an asymmetrical wave shape results. Flowing along a river will deform the flood wave: it stretches itself due to wave asymmetry. The wave rise has a higher speed than the discharging part of the wave. Hence, the hydrography influences the peak of the wave (Böhm and Deneke 1992: 181). The natural retention of the terrain depends primarily on the topographical form. The flatter a terrain is, the more water can be retained in large areas – steeper areas retain less water. A special form of retention in the terrain is snow and ice, which retains a lot of water, but comes to sudden surge in cases of rapid warming (Patt 2001: 19). Finally, these four water reservoirs retain water up to a certain extent. When one water reservoir is full, the next reservoir will be filled up. Since all the four reservoirs are overloaded, an additional runoff in form of a flood emerges.

Precipitation Floods are transformed precipitations. Thus, precipitation is the initial size to let a flood event emerge. Precipitation is a weather phenomenon, influenced by the climate. This means that the climate change has direct consequences for the emergence of a flood. According to precipitation, the Fourth Assessment Report of the Intergovernmental Panel on Climate Change (IPCC) predicts a very likely increase of the frequency of heavy rainfall events inmost areas of the world (IPCC 2007: 8). Statistics about natural catastrophes publicised by big insurance companies support this statement: they show an accumulation of damages through extreme weather phenomena in the last decades (Munich Re Group 2005: 77–82). This is a consequence of the warming of the climate system:

> [This warming is] unequivocal, as is now evident from observations of increases in global average air and ocean temperatures, widespread meeting of snow and ice, and rising global average sea level (IPCC 2007: 5).

It causes more extreme weather phenomena in the future with more dry and wet periods. In the end that would more often cause flood events with higher discharges (IPCC 2007: 8, Munich Re Group 2005: 102).

Risk increase Extreme floods are flood events that exceed the capacity of flood protection measures. The probability of floods – in particular of extreme floods – will increase in the future (Strobl and Zunic 2006: 385, 388). No matter whether the reasons for the change are naturally given or influenced by human activity (Patt 2001: 12).

River reconstructions in the past have impaired the capacity of the hydrography to cope with high water. Today, these reconstructions are considered beneficial, because they made the rivers navigable and the floodplains profitable. But indeed, these reconstructions also aggravate floods (Strobl and Zunic 2006: 389). Floods discharge faster and levels rise higher than in the past. According to retention capacities, the explanation above shows that the upper runoffs determine the character of flood events. These upper runoffs are accelerated through human activities in floodplains. In addition, the total amount of precipitation will increase, while the water masses will precipitate in rare but heavy rainfalls.

Recapitulating, even if the properties of catchment areas and the retention capacities would stay at the current level, their flood-reducing influence is confined, and aligned to lower precipitation. The increase of extreme weather phenomena will then lead to more extreme flood events in the future.

Flood Protection

In 1995, the LAWA compiled guidelines for modern flood protection. It was a reaction to the floods on the River Rhine in the 1990s (LAWA 1995: preface). These guidelines are still the standard for flood protection, and the Federal Environmental Agency (UBA) supports the LAWA guidelines (e.g. UBA 1998, 1999, 2003). The German government refers to these guidelines in their legislation (German Bundestag 2005, German Government 2005). In accordance, the guidelines framed the flood action plans of the International Commission for the Protection of the River Rhine (IKSR) (IKSR 1998), and the International Commission for the Protection of the River Elbe (IKSE) (IKSE 2003). The flood protection programmes of several federal states were also influenced by the policy paper of the LAWA: e.g. the strategies to reduce flood damages of Baden-Württemberg (Ministry for the Environment and Transport of Baden-Württemberg et al. 2003), and the concept for flood protection until 2010 of Saxony-Anhalt (Ministry for Agriculture and the Environment of Saxony-Anhalt 2003). Resulting from the River Elbe flood, the LAWA published in 2004 advice for the implementation of modern flood protection and strengthened their position (LAWA 2004).

Flood protection, declares the LAWA, contains fields of activities for technical flood protection, retention in the catchment area, and precautionary flood protection (LAWA 1995). Modern flood protection should consider all three issues in a threefold strategy (Strobl and Zunic 2006: 402). How does this modern flood protection respond to extreme floods?

Technical flood protection and extreme floods Technical flood protection is based on politically determined thresholds. These thresholds are essential for building levees, measuring dams and polders, or drawing flood risk maps (Strobl and Zunic 2006: 404). Flood risk maps document the protected space. These maps are prescribed by the EU flood protection directive (EU-directive 2007/60/EC in European Union 2007: Article 6 no. 3b, Reinhardt 2008) and by the Federal Water Act (§ 79 WHG 2010). They are the basis for measures; planners do often reason their spatial plans with the designation of areas in such flood risk maps. In this manner, technical flood protection enables intensive land uses in floodplains (LAWA 1995: 10). Water management agencies are committed to provide security up to the thresholds (LAWA 1995: 10, LAWA 2004: 21). This commitment is expressed with wordings like '*schadloser Abfluss*' (Schwendner in Sieder et al. 2007: § 28 WHG marginal 8f). The achievement of the thresholds is then the criterion for successful flood protection. Usually, hard engineering technologies are used (Moss and Monstadt 2008: 63): levees are the most popular flood protection measure (Czychowski 1998: 1171, Patt 2001: 2). Still levees are the prevailing technology of contemporary flood protection (Voigt 2005: 99). Thresholds yield levees as the predominant technology to tackle floods (Czychowski 1998: 1171, Ermer 2001: 83, Patt 2001: 2).

Manfred Voigt criticises this threshold-approached flood protection (2005). Thresholds, he explains, are approved in environmental policy; a threshold determines a maximum impact on the subject of protection, e.g. the maximal reasonable fume. If this concept was consequently applied, reasonable flood events would have to be determined. Reasonable means that the severity of an event must not exceed a determined impact on the values in the floodplain. This would indicate for flood protection that it would have to protect against rare but heavy floods, but tolerate frequent smaller inundations. Such a flood protection against extreme floods contradicts contemporary flood protection. In this manner, contemporary flood protection, based on thresholds and levees, contradicts the idea of thresholds: it protects against scarce and frequent events and fails against seldom and heavy events (Voigt 2005: 100).

In addition, levees provoke additional value accumulation in floodplains (Patt 2001: 2, Voigt 2005: 100), because landowners feel safe behind levees and become incautious (Greiving 2003: 31). This was described above as the 'clumsy floodplains'. In addition, levees raise water levels and accelerate the discharge by capturing the water body (LAWA 2004: 22). This aggravates the flood situation downstream (Strobl and Zunic 2006: 389, UBA 1998: 27–9). Levees are expensive constructions; levees influence the ecological functioning of floodplains; levees are not able to protect against extreme floods (Voigt 2005: 100). Finally, comprehensive flood protection 'can rarely be achieved by the regulation of rivers' (Moss and Monstadt 2008: 63–4). Absolute security cannot be guaranteed (Assmann 2001: 197, Voigt 2005: 100–101). The European Commission concluded in 1999 in their European Spatial Development Program (ESDP):

> Dikes and other technical flood control measures do not give a 100% guarantee of safety (ESDP 1999: article 319).

But levees are essential for flood protection. Many land uses would be impossible without levees. Levees along the whole river, however, undermine its purpose: to protect floodplains against inundations (LAWA 2004: 22). Then, levees make rivers so safe until they are unsafe.

Retention in the Catchment and Extreme Floods

> *Jeder Kubikmeter, der nicht sofort zum Abfluss kommt, ist ein Gewinn für den Wasserhaushalt, der uns auch beim Hochwasserschutz entlastet.* [Flood protection profits from each cubic metre of water, which does not discharge] (LAWA 1995: 8).

The LAWA wants to retain as much water as possible (LAWA 1995: 8). Decentralised retention measures in settled areas and in vacant land are favourite measures of the LAWA and the Federal Environmental Agency in order to avoid the emergence of floods. Land thrift is therefore promoted as well as the use of rainwater, green roofs and special agricultural cultivation methods (LAWA 1995: 8, UBA 1999: 25–7). Decentralised flood retention measures can absorb smaller and average flood events, but extreme floods cannot be controlled with such decentralised measures (Assmann 2001: 203–4).

To some extent, retention capacity can also be increased through special technical river constructions (Patt 2001: 234). For example, meandering, naturally shaped rivers have a higher retention capacity than narrowly embanked water bodies (Patt 2001: 232, UBA 1999: 25). According to extreme floods, however, the effect of such ecological river restorations is insignificant (Munich Re Group 2003a: 22).

Protecting and restoring floodplains for retention became increasingly important in the course of time, when more and more extreme floods discharged the rivers. In 1995, this claim was weak (LAWA 1995: 8–13), but in 2004, when the LAWA published policy advice for the implementation of the guidelines, the demand for providing space for the rivers became explicit (LAWA 2004: 16).

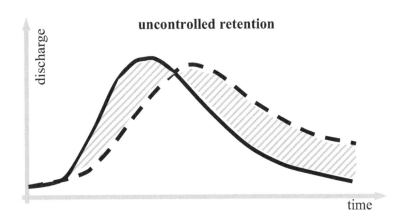

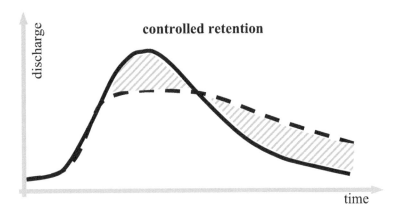

Figure 2.3   Uncontrolled and controlled retention

If forewarning and flood wave prognoses are early and precise enough, controlled retention measures can actively influence the discharge of flood waves (Patt 2001: 235, UBA 1999: 24). Retention measures can be distinguished in uncontrolled and controlled retention. Uncontrolled retention areas are often alluvial areas between the riverbank and the levee. Water can stream in and out without a barrier. Controlled retention areas have a technical construction that steers its flooding. This could probably be a small dam but it can also be a sluice. Both types of retention areas have different effects on flood waves. Figure 2.3 schematically shows the effects in comparison: the uncontrolled retention area will be inundated as soon as the water level in the river rises – so, the retention area is full when the peak arrives. Consequently, it reduces the peak only to a lower extent. The controlled retention area will be flooded when the peak arrives. It figuratively cuts the top of the wave. This explains why the controlled retention is more effective with respect to reducing the water level (Patt 2001: 235, UBA 1999: 24). In contrast, the controlled retention is less valuable from an ecological point of view, because uncontrolled retention sustains alluvial habitats (Böhm and Deneke 1992: 181). Finally, controlled retention measures are an important instrument of flood protection, because they are able to influence the runoffs essentially.

For extreme floods, however, large areas are necessary. But attempts for restoring floodplains have not been very successful (Moss and Monstadt 2008: 6), owed, as discussed earlier, to the clumsy floodplains.

Precautionary flood protection and extreme floods   Precautionary flood protection aims at preparing people and infrastructure for extreme floods and failures in existing flood protection – e.g. crevasses (LAWA 1995: 14). In the late 1990s, the political insight emerged not only to regard the inundating river, but also to consider damage reduction in the inundated settlements:

> Settlements and other uses sensitive to flooding create substantial and increasing potential for damage and loss in flood-prone areas (ESDP 1999: article 319).

The European Commission concluded this in 1999, reflecting the past major floods in Europe. The LAWA stated already in 1995 that preventing floods is impossible, but damage reduction can be achieved (LAWA 1995: 8). The ministers therefore suggested concrete precautionary measures to become more resilient to flood risk. These measures affect planning designations of floodplains, specific building standards in flood-prone areas, preparedness of behaviour during floods, and risk prevention (LAWA 1995: 14–18).

Planners should not designate new urban developments in floodplains (LAWA 1995: 14):

> *Die effektivste Begrenzung des Schadenspotenzials ist die Freihaltung hochwassergefährdeter Gebiete von Bebauung.* [The most effective measure

to reduce damage potential is keeping flood-prone areas free from new developments] (UBA 1999: 29).

Recent studies, however, show that even after the Flood Control Act from 2005, restrictions for building in floodplains do mostly refer not to existing settled areas and areas behind levees, contemporary flood protection focuses on levees and neglects imposing strict restrictions for land use and adapting existing houses to flood risk (Greiving 2006: 73).

Specific standards for building in floodplains are essential for flood protection, because it would be fatal to rely only on technical measures, and tolerate intensive uses in floodplains without preparedness to inundations (Assmann 2001: 197). How homeowners, tenants and citizens can protect their homes and prepare buildings for inundations is explained in the brochure 'Hochwasserschutzfibel' from the Federal Ministry for Transport, Building, and Urban Affairs (BMVBS 2006). It contains advice and guidelines to flood-oriented constructions, arrangements of furniture, use of cellars, flood-secure installations, safety devices for fuel tanks, etc. (UBA 1999: 30).

Becoming resilient to floods implies adapting land use and building risk adapted (Lecher et al. 2001: 474–5, Patt 2001: 57). In addition, the LAWA guidelines advice to prepare for specific behavior of people during floods. Risk awareness is often too low in levee-protected areas, which are rarely inundated (UBA 1999: 30). The 1993 flood on the River Rhine caused more than 100,000,000 DM (approximately 50,000,000 Euro) damages in Cologne; in 1995 it was circa 65,000,000 DM, despite the peak level being six cm higher than in 1993; the Federal Environmental Agency sees this as a sign for increased risk preparedness (UBA 1999: 30). But not only affected people, but also disaster management, policy, fire brigade and relief organisations must train for emergencies (LAWA 1995: 15, LAWA 2004: 27).

The LAWA advises in their guidelines to prevent financial losses and reconstruction measures after disasters e.g. with insurances (1995: 14). This advice is addressed to landowners and land users, who might be affected by a flood. The LAWA emphasises that the state is responsible only to a certain extent, and demands that potential victims of floods take responsibility and prepare for financial losses (1995: 18). However, neither the LAWA nor the Federal Environmental Agency nor flood action plans at the level of the federal states make proposals on how risk prevention might be achieved. Risk prevention remains more a political wish than an effective instrument of flood protection.

In final consequence, flood protection, as proposed by the LAWA and applied in the flood action plans, can provide protection up to a defined threshold. But it is almost unable to cope with extreme floods. This is not a new finding; rather stakeholders, academics and practitioners discuss how flood protection can be improved in the face of extreme floods (Strobl and Zunic 2006: 401–2, UBA 2003: 148).

Flood Risk Management

'Flood risk management' is the term for the new paradigm instead of 'flood protection': not protection should be promised (Grünewald 2005: 14), but the risk of flooding should be managed (Begum et al. 2007; DKKV 2003: 9, Greiving 2002, 2003, Karl and Pohl 2003: 267–70, Patt 2001: 57). This paradigm moves from the ideology that flood protection must guarantee the security of humans, flora, fauna and economic values (Boettcher 1997: 115) by defending the floods and 'keeping the water out' to an ideology of managing floods and asking citizens to 'make space for water' (Johnson and Priest 2008: 513). Making space for the rivers is the clearest leitmotiv for the new paradigm (Johnson and Priest 2008: 515). However, space for the rivers does not comprehensively explain the new paradigm; the idea of space for the rivers is not only applicable to flood protection, but also to ecological purposes – as the European Centre for River Restoration emphasises (Moss and Monstadt 2008: foreword). Flood risk management aims at providing space for the rivers, particularly in order to manage floods and flood risk.

This paradigm is reflected in many policy papers as well as in scientific publications. The European Commission, for example, uses the term flood risk management in its directive 2007/60/EC on the assessment and management of flood risks. Flood risk management should focus on prevention, protection and preparedness, therefore flood risks must be assessed and information on the risk must be provided, claims the European Commission (European Union 2007: 28). After the River Rhine floods in 1993 and 1995, the Dutch government formulated the leitmotiv of 'Ruijmte voor de Rievers' (Greiving 2002: 178).

The US Army Corps of Engineers established a national programme on flood risk management in order to integrate and synchronise activities of all involved stakeholders by providing information to the public and stakeholders, assess flood hazards, improve public awareness and become resilient (USACE et al. 2009). In the position of the German Committee on Disaster Prevention (DKKV e.V.), flood risk management is a loop of iterating processes of disaster management and prevention about flood risk management. The DKKV promotes this idea in the study about the 2002 flood event at the River Elbe (DKKV 2003). Stefan Greiving (2003) uses the term risk management in his proposal for the future organisation of flood protection authorities. So, many different meanings and concretisations of flood risk management exist. It is then not astonishing that, for some years, a journal on flood risk management has been published by Wiley-Blackwell in partnership with the Chartered Institution of Water and Environmental Management to discuss all the different standpoints and approaches to flood risk management.

In conclusion, the term 'flood risk management' stresses the ongoing paradigm shift. Policymakers, practitioners and scientists agree that another approach than traditional flood protection has to be applied for riparian landscapes. Space for the rivers is the major issue of this new approach. The following three principles of flood risk management are derived from the discussions. They explain how space for the rivers helps to cope with extreme floods.

Upstream protects downstream   Already the LAWA emphasised in 1995 that water management agencies should consider effects of measures in the floodplain with respect to downstream parties (1995: 11). The fortune at a given place on a river depends on the upstream. Almost everything flows downstream: chemicals, pollution, heated cooling water and the water itself. Upstream has the first access to the water, and thus influences the discharge downstream. The opportunity to use the water at first for drinking water, irrigation, cooling water, industrial purposes and other purposes, makes the downstream dependent on the upstream in cases of low water, when water is a scarce good. In cases of high water, the upstream decides whether to retain the water or to accelerate the discharge to get rid of the surplus water masses. Finally, neglecting some specific situations where water flows back upstream through tailback, only ships and salmons travel upstream. This dependency of the downstream on the upstream makes it necessary to involve the upstream parties in the flood protection for a certain place.

Flood risk management requires implementing the basic principle 'upstream protects downstream'. Almost every publication on modern flood protection and flood risk management claims for a consequent implementation of this principle in the whole catchment area of rivers (e.g. DKKV 2003, Grünewald 2005, LAWA 1995, Staatliches Umweltamt Krefeld 2002 and UBA 1998, 1999 and 2003). Moreover, the Commission of the European Community emphasises that:

> It is feasible and desirable to reduce the risk of adverse consequences, especially for human health and life, the environment, cultural heritage, economic activity and infrastructure associated with floods. However, measures to reduce these risks should, as far as possible, be coordinated throughout a river basin if they are to be effective (European Union 2007: 27).

The scientific discussion sustains this claim, like Klaus Ermer, who argues that reducing the flood as well as reducing the damage requires concerted concepts all over the river basin – not only for parts of it (2001: 83).

Hydraulically, the implementation of the principle to reduce the inundations is apparent. Floods have their origin spatially separated from the place of inundation. The flood in 2002 on the River Elbe emerged – considering some simplifications – mainly in the Czech Republic and produced the most harm in Germany. In 1995, some smaller events accumulated to the extreme flood on the River Rhine: the flood wave of the River Mosel overlapped with the wave of the River Rhine and produced high water levels in Cologne. So, the governmental agency responsible for flood protection at the River Rhine in Germany claims for a catchment area-wide implementation of the principle 'upstream protects downstream' (Staatliches Umweltamt Krefeld 2002: 7).

The German ministries responsible for water and the European Commission both claimed in their guidance for modern flood protection for catchment-wide implementations of the principle 'upstream protects downstream' (LAWA 1995: 1). In contrast, a water manager from the Emschergenossenschaft explained in a meeting of the Flood Competence Center in Cologne (HKC) in July 2008:

> *Bei uns ist ein Hochwasser in sechs Stunden durch.* [We discharge a flood in six hours] (Water manager of the River Emscher, 2008).

The Mayor of Bitterfeld confirmed that 'upstream protects downstream' is not important for local planning decisions, rather inundations in the individual municipality are a major concern (Mayor of Bitterfeld, interview, 1 June 2008). An officer at the ministry for building and transportation of Saxony-Anhalt explained in an interview that it is a difficult business to mobilise land for retention: retention intervenes in property rights, which is time-consuming and expensive. If a particular retention measure reduces the level of discharge for about a few centimetres, but costs a lot of money for the taxpayer, it is difficult to reason its realisation (Officer at the Ministry of Building and Transport Saxony-Anhalt in an interview in 2008).

Obviously, the principle implies all the duties for the upstream and protection to the downstream. Consequently applied, this could lead to an imbalance: the downstream expands and claims on the basis of the 'upstream protects downstream' principle for an adequate behaviour of the upstream, which implies restrictions for the use of the land in floodplains. So, what would be a viable way of treating the principle? The LAWA declared that floodplains should be used in such a way that the flood situation does not worsen for the upstream and the downstream (1995: 1). Accordingly, the principle 'upstream protects downstream' should not be used as a 'carte blanche' for the downstream, moreover it refers to a reciprocal responsibility for flood protection of the parties within a catchment area.

Integrated flood risk management   'Upstream protects downstream' is a political issue. For that reason, the LAWA claims for 'integrated flood protection' (1995: 19). In the past, flood protection was a sectoral task of water management, deserving economic interests for using land in floodplains (Assmann 2001: 198). The new paradigm acknowledges that damages can only be reduced in an integrated strategy (Grünewald 2005: 14, Kron 2003a: 92). 'Integrated' refers on the one hand to the integration of the flood issue in comprehensive spatial planning; on the other hand, it refers to the integration of all relevant stakeholders and interest groups in the planning process of protection measures. This increases acceptance of restrictions and reduces conflicts (Assmann 2001: 199).

It remains to be discussed who are the relevant stakeholders and interest groups. Wolfgang Kron, for example, proposes three relevant stakeholders: the state, which provides basis-protection with levees; citizens, who take responsibility and prepare for flooding; and insurance companies, which compensate individual losses in case of flooding (Kron 2003a: 92–5). Others depict a wider picture of relevant stakeholders and include environmental agencies, regional planning, local land use planning, disaster management, population, etc. (e.g. DKKV 2003, Grünewald 2005, Heiland 2002). Usually, regional planning plays an important role in the discussions, whereas landowners are mostly ignored. The common thought of all these ideas is that flood risk management is not a task of one stakeholder or interest group, but it is a complex task (Billé 2008: 79).

Prepare floodplains for extreme floods   Flood risk management is also the insight that an extreme flood might occur and exceed the capacity of contemporary protection measures. This is probably the most important aspect of flood risk management. Levees, as discussed earlier, can only cope with floods up to a certain threshold – extreme floods exceed these thresholds. For such extreme floods, emergency plans are required (Boettcher 1997: 124, Lecher et al. 2001: 476), because humans can steer floods only to a limited extent (Patt 2001: 11). Indeed, the water masses of extreme flood events are most often underrated, whereas the effects of technical measures are mostly overrated (Kron 2003b: 29). In addition, extreme floods exceed the horizon of experience of a generation, but a centennial flood will occur in the next 20 years with a high probability (LAWA 1995: 13–14). Sustainable preparations for extreme floods are required. Reducing flood damages requires adapting and restricting land uses along the rivers (Patt 2001: 57). It does not suffice to rely on technical flood protection measures and accumulate sensible uses in floodplains (Assmann 2001: 197).

Since flood risk is a product of probability and potential damage, the Intergovernmental Panel on Climate Change (IPCC) advises to react to both variables by mitigating the event itself and adapt to the threat in order to become more resilient in terms of damage (Fleischhauer 2004: 32–3, IPCC 2001). The LAWA advises to prepare for such extreme floods, because there will always be a flood higher than the protection capacities (1995: 13). Stakeholders must be aware of that and prepare for remaining risk (Strobl and Zunic 2006: 383). So, preparing the floodplains for extreme floods is another important principle of flood risk management.

Towards Floodplain Management

Summarising the arguments above, the following can be concluded: no matter whether climate change is naturally given or anthropogenically caused, extreme floods will increase in the future. Contemporary flood protection is incapable of tackling extreme floods. The new paradigm, flood risk management, is therefore discussed. Flood risk management is not a contradiction to flood protection, it rather proposes for another application of flood protection measures.

Finally, everyone – scientists, policymakers, practitioners – seems to know what should be done, but owed to the social construction of the clumsy floodplains (as discussed earlier), it does not happen. A wise management of flood risk instead of dogmatic claim for protection is required. Three management principles are important for extreme floods: upstream protects downstream; defending floods is an integrative task, rather than a sectoral task; and preparation for extreme floods is essential.

Flood risk management merges water policy and land policy. The European Commission stressed this relationship in the flood risk directive on the assessment and management of flood risks (European Union 2007: 28). Flood risk management copes better with extreme floods than contemporary flood protection. However, it struggles with the clumsy floodplains. I propose a comprehensive management of the land in floodplains, not a management of flood risk. This reflects the four perceptions in the four phases in floodplains. For that reason, a concept of floodplain management as an enhancement of flood risk management is proposed here: Large Areas for Temporary Emergency Retention – LATER. This concept contains the three mentioned principles and manages floodplains not only in dangerous and controllable phases, but also includes the perspectives of the inconspicuous and the profitable floodplains.

Large Areas for Temporary Emergency Retention

LATER is a concept for coping with extreme floods by making space for the rivers in a particular way. LATER stands for Large Areas for Temporary Emergency Retention. Such areas are parts of floodplains, which will be flooded in order to cut the vertex of the flood in case of extreme floods. So, LATER *s*acrifices certain upstream areas to avoid damage in the downstream area. In this manner, it prepares floodplains for extreme flooding. In the sacrifice area, adaptation of land uses is necessary, but the downstream area is not paralysed by strict restrictions through flood protection. *E duobis malis, minime eligendum est* – LATER is an institutionalised choice of the lesser evil. The retention reduces the peak discharge for the downstream party. After all, the Federal Flood Control Act explicitly declares restoration of floodplains for retention as a target of water management (German Bundestag 2005: 8 and 12). LATER takes this target seriously for extreme floods.

LATER

Technically, LATER works like usual polders. But the concept is different according to size, property regime, duration of use, purpose, and specific functionality of the retention areas.

Size: (L)arge   Decentralised retention measures like dams and polders have only small effects on extreme flood events, as well as retention by regulating lakes (Böhm and Deneke 1992: 190, Munich Re Group 2003a: 22). Flood storage basins along the main stream reduce extreme discharges most effectively (IKSR

2005: 4). In the past, enormous efforts were made: between 1995 and 2005, water management provided additional 77,000,000 m³ retention volume along the River Rhine; in sum, 213,000,000 m³ retention volume was available in 2005 (IKSR 2005: 8); until 2020, the IKSR targets at additional 296,000,000 m³ (IKSR 2001: 15). These are enormous successes for the local flood protection.

However, the here proposed sacrifice areas have to be very large. Extreme flood events even exceed the achieved retention volume. For example:

> The destructive part of the Mosel flood wave in the 1993 Christmas floods had a flow of over 2,000 m³/s and a volume of 630,000,000 m³; retaining it would have required a basin the size of Lake Constance with a water depth of 1.20 m (Kron 2003b: 29).

It would hardly be possible to realise such a retention measure along the narrow terrain along the River Mosel, but the example gives an imagination of the dimension of the water masses. In addition, the 1993 flood wave of the River Mosel was only a part of the flood, which affected the lower River Rhine in Cologne; the River Rhine itself also discharged enormous water masses.

In 2002, the flood discharge of the River Elbe reached at its peak 5,000 m³/s in Usti nad Labem (Czech Republic) and produced an eight-metre high wave in Dresden. This illustrates the size of the retention of LATER. The *Havelpolder* provided a retention volume of 75,000,000 m³. The largest area, flooded by a crevasse between Torgau and Dessau, spanned 30 km. Between 17 August 2002 and 19 August 2002, 300,000,000 m³ of water streamed in an area of 193.9 km² (Engel et al. 2002: 17). The water masses between Torgau and Dessau covered an area approximately as large as the settled area of Berlin, or the double size of Dresden. This size by far exceeds usual retention measures.

LATER aims at retaining extreme floods. Extreme floods inundate extreme large areas. Areas in a much bigger size than usual polders are required. But LATER is no substitute for such polders, it is just designed for another purpose: in case of calamity, these areas provide an additional retention volume in order to minimise damage.

Allocation: (A)reas in risk alliances   Retention areas have to be allocated in hydraulically suitable places. Information about the probability and intensity of floods in the catchment area is essential, as well as detailed knowledge of hydraulics. Contemporary, retention polders are realised after experts prepared feasibility studies for different alternatives. These studies include effects of the measure on the discharge, ecological and environmental issues, effects on groundwater, contamination of soil, etc. Often, already these studies stoke objections of landowners against a measure. In the end, the allocations of polders result from political considerations and negotiations.

LATER aims at reducing the damage of extreme floods for particular downstream parties. The difference to the traditional polders is that landowners and their immobile values are in the very centre of discussions, rather than political power relations. Risk alliances result from mutual agreements between affected landowners. The state or municipal agencies might also be involved to provide information or mediate between the parties, but the allocation decision depends on the risk alliances. The essential task of public agencies is to initiate and support the risk alliances. This issue, however, will be discussed later, after some other issues of LATER are explained in the next paragraphs.

Duration of use: (T)emporary   Land for retention polders is usually designated by a formal plan approval (*Planfeststellungsverfahren*). This designation imposes several restrictions for land use. Often, only extensive farming is allowed and building is impossible in these areas (for example in the *Havelpolder*). In addition, polders are flooded quite often, which technically restricts land uses. A usual dispute emerges between farmers and environmental interest groups. Farmers want to use the land as profitably as possible, environmental activists claim for frequent flooding of the land in order to sustain alluvial habitats. These restrictions on land use endure as long as the polder exists, according to a generation this is almost forever. No matter whether the flood takes place in the next few months or 100 years and lasts only a few days or even weeks.

Conflicts with landowners are crucial for the realisation of polders. Authorities take landowners as serious obstacles for implementation. If previous studies estimate too heavy conflicts with landowners, authorities withdraw from certain measures, as happened, for example, with a draft for a large version of the polder Cologne-Langel (Bezirksregierung Cologne 2003). The polder was planned in 2003 for a retention volume of about 4,525,000 m³ (Bezirksregierung Cologne 2003: 45), which is a big measure indeed, but compared with the numbers mentioned earlier (like 630,000,000 m³ in 1993 at the River Mosel) it can only be a small contribution to space for the rivers. Even though, the biggest chapter in the formal plan approval for this particular measure was about property-related objections of individual landowners: 165 private landowners raised objections against the polder Cologne-Langel (Bezirksregierung Cologne 2003: 81–99). Citizens sued against the formal plan decree. The diploma thesis from Frank Gerdes (2008) confirmed that farmers are often anxious about economic impacts. It came out in this study and other student projects that conflicts with landowners are typical for retention measures that require formerly privately owned land. The availability of land, however, is crucial for the implementation of retention measures.

LATER measures are much bigger than usual polder measures. Restrictions like those mentioned above would be serious obstacles for LATER; in particular, when these areas are needed very seldom. Over the lifetime of a whole generation, the emergency retention might not even be needed. In advance of the 1993 flood, Germany experienced such a dry period. LATER might also be needed two or three times in a short period: the River Rhine discharged extreme floods within two years; the River Elbe had three major events in the period between 2002 and 2006. Finally, LATER is not an everyday land use, it is rather a temporary land use.

LATER measures can be used for other purposes than retention as well. Settlement or business locations are possible in these areas as long as they are adapted for a controlled flooding by technical measures. The required techniques exist already. Mitigation measures at the level of households can effectively limit flood damage during floods (Botzen et al. 2009: 25). In his study about the flood events of 2002, Uwe Grünewald summarises methods for protecting existing objects in flood-prone areas by constructional protection measures (DKKV 2003). Such measures are the most effective in areas with low but frequent inundation depth. The experiences in the towns often affected by a flood, like Cologne, show how permanent or mobile barriers can protect houses. To improve the stability against uplift, water pressure and flow forces, the buildings can be anchored and weighted. Backwater valves protect against water entering through the sewage system. 'Black wells' and 'white wells' can protect the foundation of buildings. A 'black well' uses special foil to defend against the water; a 'white well' consists of a special kind of low water permeable concrete. Such measures should defend from infiltration of water as long as the constructional stability of the building is guaranteed. If the building becomes unstable due to the external water pressure in the groundwater, homeowners are advised to admit the inundation of the house. In advance, the damage can be reduced by a flood-adapted use of the building – e.g. by putting valuable facilities and technical infrastructure in the higher floors. Furthermore, the cladding can be constructed as water resistant with special paints and building materials. A survey of the GeoForschungsZentrum Potsdam and the Deutsche Rück Reinsurance Company pointed out that a flood-adapted use reduced the damage in the flood event in 2002 in the catchment area of the River Elbe: damages of personal belongings by between 13 and 15 per cent, and structural damages in buildings by between 8 and 9 per cent. Often, the landowners implement the constructional protection measures after a flood event (DKKV 2003: 18–20).

LATER allows almost normal land uses within the retention areas. Due to the inherent threat, nuclear power stations or chemical industries should in fact not be allocated within the submergible land (though in particular these land uses are already located in riparian landscapes). The idea of retention areas as temporary land use as here proposed aims at protecting objects instead of protecting areas. This ideology respects the profitability of floodplains. Retention of floods, which occur once in every 50 years, must not paralyse areas as large as LATER and as fertile and valuable as riparian land.

Purpose: (E)mergency   However, in case of an extreme flood, the downstream must be able to retain the flood wave on the upstream land. Barrie Needham explains that, probably:

> others besides the owner or user of a plot of land have an interest in how the
> owner or user uses that plot (2006: 9).

In the case of LATER, the downstream party has an interest in the use of upstream land. They indeed want to sacrifice the upstream land for their own sake. Past flood events depict what this can cause in practice: in 2002, an extreme flood occurred on the River Elbe, in 2006, another flood discharged the Elbe. In 2002, the small town Hitzacker in Lower Saxony was not inundated – despite there have been no appropriate levees or other protection schemes. In 2006, which was a smaller flood event, Hitzacker was inundated. This is a consequence of the flood protection policy in the upstream areas: after damages in 2002 in Dresden and along the whole River Elbe in Saxony and Saxony-Anhalt, which is upstream to Hitzacker, the levees were rebuilt and strengthened. In consequence, they defended the water in 2006. In 2002, 12 larger crevasses (Engel et al. 2002: 17) caused damage in large areas in Saxony and Saxony-Anhalt. In 2006, Hitzacker received the full amount of water at a high peak. In other words, the unintended inundations in 2002 in Saxony and Saxony-Anhalt worked like controlled retention measures (IKSE 2003: 24), and so protected Hitzacker. Even though that this is a cynical perspective, the historical building 'Dresdner Zwinger' served Hitzacker in 2002 as a retention area. A similar situation can be expected for the next River Oder flood: in 2010, large areas in Poland (upstream) have been inundated, whereas German floodplains along the River Oder remained dry. What happens if Poland decides to improve their levees substantially after the disaster (which is, according to the social construction of floodplains, probable)?

Obviously, without undervaluing the harm in Hitzacker, the described case was far from the most efficient allocation of retention areas. But acknowledging that not all areas along rivers can be protected against extreme floods, at least the most important areas should be protected in order to reduce the flood damage. LATER identifies and provides sacrifice areas. They have one important function: to retain water masses for the sake of valuable downstream areas. Several measures within the sacrifice areas can contribute to further reduction of damages: specific building standards, restrictions for agriculture (e.g. by using flood-resilient crops), and flood forewarning to initiate evacuation of humans, animals and material (Vischer and Huber 1993: 284–6).

Functionality: controlled flood (R)eduction   LATER protects through controlled retention. Technically, the probability of an inundation at a given place can be reduced through defending the water with levees and similar constructional measures like bypasses, or flood risk can be reduced by lowering the maximum discharge level through retention measures upstream (Lecher et al. 2001: 474–5). One of the most effective ways to reduce flood damages is indeed retaining the water upstream (Cooley 2006, Strobl and Zunic 2006: 416). Theresia Petrow and her colleagues propose in case of extreme floods to use open spaces along rivers to reduce the maximum level of a flood wave and mitigate with this measure the damage downstream: 'these retention areas can be used for emergency discharge' (Petrow et al. 2006: 718). Such a controlled retention reduces the maximum discharge for the original inundated area to an acceptable level (Vischer and

Huber 1993: 282). However, the effect of a controlled retention diminishes due to distance (IKSR 1999: 6).

Cutting the flood wave at its vertex requires precise forewarning and controllability of the polders (Strobl and Zunic 2006: 420). Such controlled retention is only possible if the area of interest for protection is not to too far from the retention (IKSR 1999: 6). LATER is not applicable for flash floods, where prediction is difficult, imprecise and short. This makes controlled flooding almost impossible (Strobl and Zunic 2006: 392). But LATER is applicable for large catchments (Strobl and Zunic 2006: 396). Whether LATER *is* suitable for a river depends very much on the specifics of the catchment – its size, shape and relief.

To increase the effect of the retention measures, close to the retention area the flow velocity will be decelerated; the tailback of the water facilitates the filling of the retention polder. The water can then be discharged later; respectively it seeps or evaporates (Vischer and Huber 1993: 289). So, the technologies for LATER are already available.

Property-focused Risk Alliances

The here proposed concept requires not only technological preconditions, but risk alliances between upstream and downstream parties have to be established. Such risk alliances are based on a mutual agreement between the provider of the retention, and the beneficiary of the retention. This risk alliance distinguishes the concept of LATER from the Dutch concept of calamity polders (*noodoverloop-gebieden*), or the concept of emergency spillways (*Notentlastungen*) by the IKSR (IKSR 2002: 38). These risk alliances are property-focused approaches to retention areas.

Noodoverloop-gebieden *and* Notentlastungen   The Dutch government launched in 2000 the policy directive 'Room for the River', which ought to prepare the country for the future (Roth and Warner 2007: 519). By a controlled flooding of sparsely populated areas, densely populated areas with strategic economic functions should be protected. The Dutch call this concept *noodoverloop-gebieden*, calamity polders. A committee proposed three possible areas: the Ooijpolder near Nijmegen, the Rijnstrangen in Gelderland and the Beersche Overlaat in Brabant (Ministrie van Verkeer an Waterstaat 2009). Calamity polders were seen as last-ditch emergency measures. Finally, owing to strong resistance of civic action groups, the Dutch calamity polders are only moderately successfully implemented (Roth and Warner 2007: 520–21).

The International Commission for the Protection of the River Rhine describes *Notentlastungen* as an emergency measure of disaster management. The Commission proposed to flood certain areas with less potential damage in order to avoid an uncontrolled inundation of large areas with enormous damages. Existing linear structures like roads, old embankments and levees, canals etc. serve as 'second line of defense' – *zweite Verteidigungslinie*. The IKSR emphasises that the land uses are not restricted in these areas, because the 'normal or statutory' level of protection is provided (IKSR 2002: 38–9). However, the idea of *Notentlastungen* is not very popular and caused political resistance. Although the IKSR publishes many papers on floods, *Notentlastungen* are rarely mentioned.

LATER is similar to the Dutch *noodoverloop-gebieden* and the German *Notentlastungen* from the technical point of view. Both ideas sacrifice certain areas in order to reduce the harm in the whole catchment. The difference between LATER and the two ideas presented is the approach to land and in particular to property. In the Dutch and the German proposal, land is needed as an essential. Property is regarded as an obstacle for the implementation. LATER puts property in the centre of interest. LATER serves for proprietors as an instrument to gain and save money. This is the property-focused approach to land in floodplains.

Property focus   The particularity of the concept of LATER is that upstream and downstream establish a risk alliance. In advance of extreme floods, both agree on terms for controlled inundation of the upstream. Such terms may stipulate payments from the downstream to the upstream for the right to retain the next extreme flood wave on the land of the upstream party. The downstream landowners save money in the next extreme flood; the upstream landowners gain money from providing retention volume on their land. This is the property-focused approach of LATER: landowners make a mutual arrangement on the allocation of the next extreme flood. Interviews confirmed that indeed many landowners are willing to pay for additional flood protection, as long as the effect for their property is guaranteed (interviews with landowners).

Technical and administrative proceedings are postponed in the concept of LATER; land management is first. If constructions like spillways or intake structures are necessary in the levees, all relevant regulations need to be taken into account: formal plan approval (*Planfeststellungsverfahren*), environmental impact assessment and special regional planning procedures (*Raumordnungsverfahren*) (Spieth 2008: § 31 WHG marginal 45). The IKSR assumes that regional planning designations are not affected by emergency retention measures, as long as the existing level of protection is not lowered (IKSR 2002: 39). Building water barriers in the hinterland or using existing structures like railway or street dams in order to avoid inundations of valuable areas are subject to the same administrative proceedings as for summer dikes. However, these proceedings are postponed; the property-focused approach mobilises the land first, but it does not release authorities from their administratively necessary responsibilities and proceedings.

Inundation easements   This is an essential difference to the usual approach of realising retention measures: local and regional authorities usually plan and implement polders. Once the respective area is designated for retention, water management intends to become the proprietor of the respective land. The sites in retention areas are most often privately owned; private landowners, though, are usually seen as an obstacle for the implementation of the plan (Reinhardt

2003). For that reason, responsible authorities order feasibility studies to evaluate planned retention measures. These studies ought to prove the public necessity of each measure, in order to justify interferences with private property of floodplains (officer at the Ministry of Building and Transport Saxony-Anhalt in an interview in 2008), and justify compulsory purchase, if necessary. However, practitioners in fact almost never use compulsory purchase to mobilise the land for retention; prevalent methods are freehand purchase, barter agreements on sites and land readjustment (UBA 2003: 121). No matter how water managers get the land, they typically strive for full ownership. My own research during my diploma thesis in Saxony-Anhalt, and results from student projects in the south-west of Germany, along the River Lippe, and in Cologne confirmed this widespread practice – except in one project in the Pfalz, where a smaller measure is realised on the legal basis of a long-lasting building lease (*Erbbaurecht*).

The size of LATER is extraordinary large. A large number of landowners will be involved. Traditional land management approaches for retention sites, namely freehand purchase, barter agreements on sites, or land readjustment will not work; the number of objections, the required money and time would paralyse such big projects.

LATER pursues another approach. Landowners are not seen as an obstacle for the implementation of retention measures, but rather landowners are in the centre of the implementation. Landowners remain proprietors of their land. As property is a bundle of rights (Cooter and Ulen 2004: 77), LATER needs only one stick of this bundle: the right to inundate the respective sites in case of an extreme flood. Such a stick could be an easement, a lease or a leasehold interest (Needham 2006: 95). A leasehold interest allows using the land for a certain period, a lease allows another party to use a part of the land, and an easement allows the usage of the land in one particular way. The appropriate right for LATER is an easement. The landowner grants the downstream party the right to inundate the land probably one time in 50 years by approximately 2 metres (or by what is in fact hydraulically required). This right is registered in the cadastre. The inundation easement is like a way-leave, which is not like a leave to cross a piece of land in a daily-turn by a car, but like crossing a piece of land rarely by a steamroller. Initiating these risk alliances is an essential task for the implementation of LATER.

Floodplain Management by LATER

As mentioned earlier, LATER is a concept for managing floodplains. It is a particular kind of flood risk management. Three principles must be matched to manage the risk of extreme floods: upstream must protect downstream, the management must be integrated, and it must prepare floodplains for extreme events.

It has already been explained that LATER implements the principle 'upstream protects downstream' within risk alliances. The upstream and the downstream party become personalised with the concept of LATER, they are no longer anonymous actors. This is important to initiate actions of the stakeholders: LATER creates concernment, which is important for implementation of such measures (officer at the Ministry of Building and Transport Saxony-Anhalt in an interview in 2008). The risk alliances necessitate the integration of all relevant stakeholders, whereas the most important aspect of LATER is the integration of upstream and downstream landowners. LATER aims at extreme floods, but due to the inundation easements and approach that retention is a temporary land use, which sets incentives for adaptive measures in floodplains.

LATER not only steers risk, but it also manages the land use in the whole floodplain: due to the inundation easements it becomes more difficult to settle in the retention areas. These are most probably rural areas. Instead, the additional protection in the retention areas sets incentives to build in these areas rather than in rural floodplains. Housing markets often reflect the risk of natural disasters (Schwarze and Wagner 2004: 160). The payments to the upstream for LATER internalise the flood risk and it becomes more attractive to settle in less flood-prone areas. In this manner, LATER manages not only flood risk but influences the land use in floodplains in general.

Economics of LATER

Floodplain management by LATER aims at reducing damages. Efficiency is the most important measure to assess the success of LATER; namely to what extent economic losses have been avoided. However, although 'efficiency is always relevant to policymaking' (Cooter and Ulen 2004: 4):

> still, efficiency is used too often as an excuse for avoiding the issue of justice and fairness (Davy 1997: 254).

Barrie Needham and Benjamin Davy agree that land policy should achieve fairness of distributional effects besides economic efficiency (Davy 2005: 117–19, Needham 2006: 139):

> Conventional policy analysts of course will argue that fairness, unlike economic efficiency, is a value-laden concept and therefore has no place in the dispassionate business that they are engaged in [... But] there is nothing neutral or dispassionate about the criterion of economic efficiency (Ellis and Thompson 1997: 212).

Indeed, efficiency pursues an inherent concept of fairness: who puts most in should get most out. So, achieving efficiency is not independent from fairness. Barrie Needham emphasises that just distribution should also include sustainability:

> Does 'our generation gain at the expense of future generations' (2006: 139).

For these reasons, the following discussion on efficiency of LATER considers distributional effects when considering gains and losses from LATER in the flooding game as well as implications for the future.

Efficiency of LATER

LATER makes landowners poorer or richer:

> by affecting the value of the rights on their land ... those financial consequences for individuals can add up to very great economic consequences for the society (Needham 2006: 3).

Flood protection – a scarce good   Flood protection by levees provides profitable land use in floodplains. The product of flood protection is security. As mentioned earlier, such protection is limited, in particular with respect to extreme floods. Economists call such limited goods scarce. Protection is scarce since it can only be increased up to a production possibilities frontier. The production possibilities frontier shows possible alternative combinations of using the available factors of production for two goods that an economy can produce (Cooter and Ulen 2004: 183). Once this frontier is reached, each portion of the resource is in use (or, with respect to depleting resources, was used), which means that no resources are wasted. At this frontier, additional and higher levees will not improve safety. On the contrary, risk actually increases, as discussed earlier (contemporary flood protection is not able to cope with extreme floods). Finally, a paradox emerges. Namely, additional protection measures will make rivers unsafe.

In the system of only one upstream and one downstream party, the upstream yields flood protection by heightening the levees and accelerating the discharge. The downstream yields flood protection by retaining water on the land of the upstream (and of course, the downstream could forward the flood to the next downstream, but then the original downstream becomes an upstream; for that reason, we regard a system of only two parties along a river). The allocative choice is either to produce protection for upstream or to protect downstream. Figure 2.4 shows the production possibilities frontier for flood protection. At the point A in the figure, protection capacity is wasted, because levees can be built for the sake of the upstream, or retention areas can be designated upstream without imposing each party. At point B, however, production of one good can only be enhanced by impairing the production of the other good (Cooter and Ulen 2004: 16–17, HdWW Bd. 2 1988: 524). The line represents Pareto-efficient allocations of the flood protection.

What happens in a Pareto-efficient situation if one party should be better protected for some reasons? Assume the upstream should be better protected (because it is a valuable city). This implies to heighten the levees upstream. In consequence, the protection level downstream will be reduced, because the flood wave will be accelerated and higher water levels discharge downstream. So, the

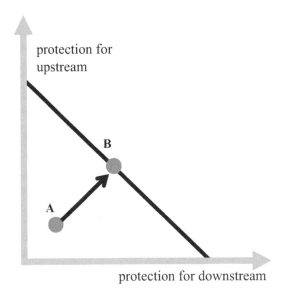

Figure 2.4   Production possibilities frontier

improvement upstream produces an impact on the protection downstream. Vice versa, assume the protection of the downstream should be improved. A heightening of the levees would cause effects further downstream, and thus, outside the two-party system. In the two-party scheme, the downstream can only be better off by retaining the water upstream. This, however, reduces the protection upstream, because upstream will certainly be inundated in cases of flooding. Finally, improving the protection for one party, burdens another party. It is impossible to improve the protection of both parties – in other words, flood protection is scarce and it is already allocated Pareto-efficiently.

A decision has to be made whether to protect the upstream and sacrifice the downstream, or to protect the downstream and sacrifice the upstream. Actually, the latter is the case of flood protection today, but nobody asks the downstream parties if they can cope with the additional impact, they just forward the flood to the next downstream. The downstream at the end of the river suffers losses, as the case of Hitzacker showed. Who should be protected?

This situation is similar to the 'meat or crops' problem that Ronald Coase was concerned with in his article about social costs:

> If it is inevitable that some cattle will stray, an increase in the supply of meat can only be obtained at the expense of a decrease in the supply of crops. Nature of the choice is clear: meat or crops. What answer should be given is, of course, not clear unless we know the value of what is obtained as well as the value of what is sacrificed to obtain it (Coase 1960: 2).

Values of both goods must be assessed in order to decide which good is preferable. How can these values be evaluated? Flood protection has no fair market value like meat or crops. Flood protection has to reduce damage. Hence, the preferred damage is the benefit of flood protection.

Prevented damage – benefits of flood protection   'Damage: injury or harm that reduces value, usefulness, etc.' (Random House Webster's Dictionary 1993). The Merriam Webster Online Dictionary determines 'damage: loss or harm resulting from injury to person, property, or reputation' (www.merriam-webster.com). Stefan Greiving determines damage as destruction, reduction or impairment of concrete or abstract values by human activities or a natural occurrence (2002: 28). Obviously, different definitions of damage exist. All have in common that valuable things are to be reduced. Valuable things require an evaluating subject. This evaluation is, however, a normative process, which depends on the perception of the evaluating subject. Even the statement that a certain event produces damage is normative (Greiving 2002: 31). Stefan Greiving in this context speaks of the power of definition: who defines damage, influences what will be protected and what will be sacrificed (2002: 32). For that reason, the definition of the term damage is a crucial issue. All kind of proceedings are proposed to cope with this burden (Berg et al. 1994: 11–17).

The concept of LATER also requires a thoughtful definition of damage. Based on the evaluation of damage – or actually prevented damage – allocative and distributive decisions will be made. For that reason, it is very important to have a definition of damage, which enables a quantitative comparison of different kinds of damage. What are valuable things, and how does a reduction of its value take place?

The latter question refers to the damage-producing event. Inundations seem to be such events for LATER. The term inundation comes from the Latin term 'unda', which means wave. Thus, inundations describe situations where objects are submerged under or in waves (Doe 2006: 2). In some cases, this inundation can have positive effects, e.g. for alluvial forests. Sometimes, however, the effects of inundations have negative consequences for the inundated objects. But an inundation is not per se relevant for LATER, although an inundation is a damage-producing event. Only inundations caused by river floods are of interest.

According to valuable things, it is more difficult. What kind of valuable things could be reduced by a flood-caused inundation? To answer this question, it is helpful to build up categories of different types of reductions of valuable things. Stefan Greiving (2002) uses three categories of damage: effective damage, eventual damage and damage of clearing the catastrophe. The first is a loss of real values like the detraction of an entitlement, a right, or such an interest of individuals or communities protected by law. Eventual damage means losses of real or pretended opportunities. In principle, these are the opportunity costs of the inundation. The effort of cleaning up and rebuilding is the third category. These three are a first approximation to the variety of negative effects of inundations. A study of the German Environmental Agency about flood protection identified three other categories of damage: damage to human beings, economic loss and environmental damage (UBA 1999: 13). The former are primarily the dead and injured humans, but also the evacuated people. Economic loss includes direct losses of the private, the commercial, the industrial and the public sector as well as losses of crop and production. Costs for disaster control, costs for rebuilding and public financial support are contained in this category too. Some of these costs are covered by insurance. The third category, the environmental damages, is difficult to grasp. It encompasses, for instance, toxic depositions. In media reports, this category plays a lower role, because the events are mostly dramatic for humans and economics (UBA 1999: 7). These categories operationalise to some extent the real damage from Stefan Greiving's categorisation. Examples for concrete economic losses are impairments of land values, which are the concern of Matthias Rötzsch (2005). These losses can be technical, mercantile and operational impairments. The technical impairment expresses the real capacity of the object to use in the usual way. A technical loss of value does not emerge per se in floodplains; it emerges when an object is for sale. Technical losses have to be appraised in each individual case. After restoring the object, technical losses do not exist any longer. It can be appraised by the costs of restoring. Nevertheless, because of possible not-apparent disadvantages there can be a mercantile loss. Mercantile impairments exist – in opposite to the technical – without selling an object. They emerge though an object is restored after an inundation. The public has the suspicion of further disadvantages, which influence the selling price. Objects in a floodplain in general have a mercantile loss of value. It emerges by the risk of future inundation. Nonetheless, objects in riparian landscapes are still relatively expensive. Operational impairment, finally, emerges due to lost output of local economy (Rötzsch 2005). This example shows the broad variety of kinds of damages by inundations.

Since the purpose of the damage definition is to compare damages in different areas with each other to decide on allocation and distribution of flood protection measures, only objectively measurable types of damage can be taken into account. These types do not include all kinds of damage, but it is assumed that the measurable types of damage can be used as an indicator to compare the damage in different areas. Now, these types have to be weighted up against each other. The most appropriate method to measure damage is the monetary valuation (Kron 2003b). Nevertheless, dead and injured people have to be counted separately (Kron 2003b). Cultural goods and objects with a subjective value cannot be evaluated monetarily. Furthermore, it is difficult to evaluate environmental damage. Thus, not only the quantitative dimension of damage but also the qualitative dimension has to be regarded. In some cases, this can lead to difficult and political problematic assessments of damage. Some effects, for example environmental damage, cannot be measured monetarily, but at least the negative effects can be related qualitatively to each other. If the trend is similar to the monetarily measurable damage, monetarily measurable damage can be used as an indicator for damage as a whole. Monetary valuation is a viable method to show the change – namely the reduction or the increase – of damage.

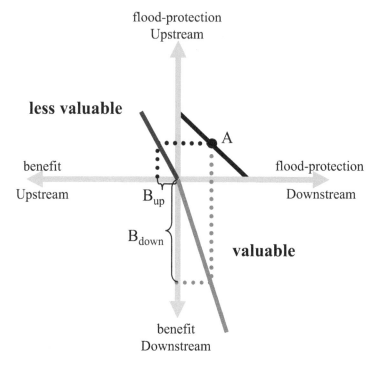

Figure 2.5   Flood protection and benefit

Asymmetrical benefits   If the benefit resulting from protection upstream is equal to the benefit from protection downstream, it does not matter who should be protected. Not efficiency but another indicator to allocate flood protection has to be applied. However, in most situations along the rivers, we are dealing with asymmetric benefits from flood protection. As determined above, damage is a loss of benefit. Consequently, flood protection is a benefit-securing measure. The value of this benefit increase equates to the value of the flood protection. Different allocations of flood protection must be compared in relation to the value of avoided damage.

Figure 2.5 shows an example of a valuable downstream area and a less valuable upstream area. The diagram shows Figure 2.4, extended by two axes, which depict the benefit from flood protection. From the origin, four independent axes emerge. The axes upwards to the origin and to the right show units of flood protection, as depicted in Figure 2.4. The axes to the left and downwards relate the units of flood protection to its benefit for the respective parties. None of the axes can become negative. The depicted case shows a particular situation: upstream is less valuable than downstream.

The upper and the right axes draw units of flood protection, as in the production possibilities frontier. For a better understanding, 'units of flood protection' can be seen as levees for the upstream and cubic meters of upstream-allocated retention for the sake of the downstream, respectively. The lower and the left axes in the diagram show the benefit of these protection units for each party. Each unit of flood protection produces benefit for the particular party. For example, a certain metre-high dike upstream protects a certain amount of values upstream. The same amount of protection downstream (not by levees but by upstream retention) protects a certain amount of values downstream.

As shown above, every protection unit upstream has an impact on the maximal possible protection downstream and vice versa. Every efficient allocation of protection units produces a certain benefit for both, the upstream and the downstream. Nonetheless, they do not achieve the same benefit. If the upstream area will get the best protection, which means to allocate the whole flood protection at the upstream, the maximum benefit is the value of the protected area. Perhaps it is only cheap farmland and some sprawled houses. In this case, the downstream area has to suffer the flood. On the other side, if the downstream will be protected best – assume it is very valuable, like a big city with high-density, expensive infrastructure, and important industrial and commercial areas – we achieve a high benefit from the same level of protection but differently allocated. Thus, every allocation of flood protection would produce certain benefits. Point A in Figure 2.5 refers to the current situation: flood protection is distributed between the upstream and the downstream according to the principle of parity. Each party – the upstream and the downstream – can each receive the same flood protection. The benefit each gains are marked as B_{up} and B_{down}.

Improving efficiency means to reallocate A so that the summed benefit of upstream and downstream is maximised. In Figure 2.6 we can see the results of different distributions of flood protection. The diagram relates the benefits of both parties with each other. Figure 2.6 is an extension of Figure 2.5. In case B, the upstream party obtains all the flood protection. The resulting benefit distribution is $B_{benefit}$. For the case of C, the benefit distribution is $C_{benefit}$. In consequence, we can enhance the summarised benefit in the case of A – that is the 'flood protection of parity'. By taking protection from the upstream and giving it to the downstream, the upstream loses only little benefit, but the downstream gains a lot of benefit. In final consequence, the summarised benefit is increased.

For the allocation of flood protection units, Pareto-efficiency is achieved in every distribution at the graph between B and C. According to the benefit, efficiency depends not only on the allocation but also on distribution. Pareto-efficiency does not lead to the most efficient solution. Economists therefore developed a 'potential Pareto improvement' or 'Kaldor-Hicks efficiency' (Cooter and Ulen 2004: 48). Accordingly, an allocation is efficient if the net benefit of the advantaged people exceeds the net loss of the disadvantaged people (Cooter and Ulen 2004: 48). This implies that LATER is an efficient alternative management to contemporary flood protection in cases where the Kaldor-Hicks criterion is met. In other words, the loss by the targeted inundation in the upstream area has to be less than the

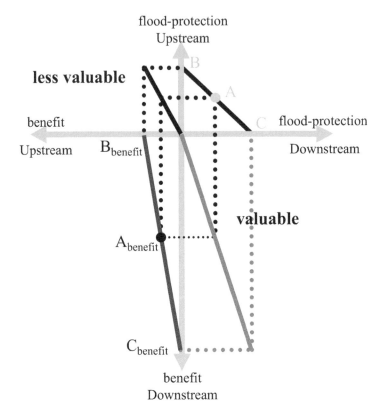

Figure 2.6   Comparing benefits of upstream and downstream

benefit created by inundating downstream. In this situation, a net benefit increase is achievable through redistribution of gains and losses.

Gains and Losses from LATER

Who should pay? Need those pay for LATER, who occupied retention volume upstream by building up levees; or should those pay, who expose their values to the risk of flooding, although the upstream has already built high levees? Should the economically most powerful parties support the weak, or should the most effective party profit most from LATER? Must the burdens and disburdens be distributed equally in the floodplain?

Donald Scherer introduces upstream–downstream relations in terms of externalities: an action upstream has, at a given point in time and space, a significant effect downstream. It can be later or distant (Scherer 1990: 19). It would be an easy way to escape discussions on justice: the victim of the externality has to be compensated by the polluter, or the polluter is obliged to care for precautionary measures, respectively the community compensates the victims. This treatment of the problem has the advantage that well-established principles of internalisation are applicable.

Principles of internalisation   Three basic internalisation principles exist. The precautionary principle aims at measures in advance of an impact. The polluter-pays principle is a principle to regulate the impact on one hand, and it sets incentives to avoid impacts. The third, the principle of the common burden is usually for cases where a polluter cannot be identified (Hartkopf and Bohne 1983: 86).

The first of the three principles is the precautionary principle. It sustains protection of nature in advance of impacts on environmental goods. Environmental goods can be used directly by using resources, but also indirectly by impact on the goods through human activity. The 'precautionary principle' aims at the latter, namely the reduction of intensity of impacts on the environment. There are two kinds of such impacts: impacts because of correct use of environmental goods, and impacts from an incorrect use of environmental goods. The difference is that in the first case there is certainty about the fact, that there is an impact, but uncertainty about the consequences. In the second case, there is insecurity about the probability and the dimension of the impact. Both cases require a comprehensive analysis of the potential impacts. Identifiable and intolerable impacts and further impacts (including currently unlikely impacts) have to be defined. In cases of correct use of environmental goods, identifiable and intolerable impacts have to be avoided; other impacts have to be decreased. In cases of an incorrect use of environmental goods, identifiable and intolerable impacts have to be avoided too, and their consequences have to be decreased, but further impacts have to be accepted (Hartkopf and Bohne 1983: 91–108). Finally, the analysis of impacts is essential to apply the precautionary principle. In addition, it is only applicable in advance of impacts.

As the name implies the polluter-pays principle stands for liability of generators of externalities. Subjects responsible for pollution or other impacts on the environment have to pay for the impact they produce. The idea focuses on the fact that an impact on the environment – a public good – burdens the public. The impact will be regarded as the polluter's costs. The polluter has to calculate costs for reducing impact (e.g. by ecological production procedure) and costs for removal of the externalities. The first type of cost is influenced by the precautionary principle, the polluter-pays principle aims additionally at the second type of cost (Hartkopf and Bohne 1983: 108–13).

If no polluter can be identified, the public has to care for the removal of pollution and cope with the impact on the environment. This is the principle of the public burden. It is ineffective from several perspectives. First, if the public pays for the reduction of an impact, the money is not available for other common tasks. Second, the polluter is not induced for impact avoidance. Therefore, this principle is only an exceptional solution for example for sudden or extreme heavy events (Hartkopf and Bohne 1983: 108–13).

These three internalisation principles differ in their approach to distribution. Indeed, the precautionary principle makes no clear statement to distribution, but the two others are predominantly concerned with distributive aspects. Obviously, for the polluter-pays principle the polluters are addressed. The principle of the common burden bears on the public – not the polluters. The principle of common burden is ineffective according to the reduction of externalities (Hartkopf and Bohne 1983: 108). Nevertheless, it is the usual principle in current flood protection. For LATER a working internalisation is essential. Thus, the polluter-pays principle is preferable.

Pollution, polluters and victims The polluter-pays principle requires an identifiable polluter. A polluter is someone who might cause pollution in the future, pollutes already, or polluted in the past (Beder 2006: 36). A victim of pollution suffers the disadvantages from the externality. Often it is not apparent who the polluter is and who the victim of an externality is.

Imagine victims have exposed themselves to the pollution. This might be irrational, but it could happen in order to receive payments for the damage or because of a lack of knowledge (probably, neither the victims nor the polluters know about the negative consequences of their allocation decision). Claiming payments from the polluter would then be unfair. In this particular situation, the polluter and the victims are difficult to distinguish (Baumol and Oates 1988: 211).

Flood protection is such a situation. Due to time, people settled in riparian landscapes and thereby exposed themselves to inundations. Upstream parties cannot be made responsible for the decision of the downstream to expose themselves to the river. However, people in the floodplains are neither responsible for the damage they produced, nor the additional retention they require upstream for protectoral reasons. They decided under conditions of lacking knowledge about the future risk development. Finally, both parties, upstream and downstream, are simultaneously both polluters and victims.

A further problem in upstream–downstream situations is that no single person could be made responsible and liable:

> It can easily happen that many agents upstream will cause effects none of them would have individually caused (Scherer 1990: 20).

In many upstream–downstream situations, the reason for externalities is a result of the cumulated behaviour of many individuals. The problem with individual insignificance is that the individuals do not act as members of a community. Therefore, Donald Scherer proposes to make individuals aware of being part of a community, so that the well-being of the community becomes a weight for the individual decision-making (Scherer 1990: 33–4).

Who should pay for the externality (damage through floods)? Two different schemes could be applied: either Arthur Pigou's proposal of a tax to internalise externalities, or Ronald Coase's position of promoting negotiated agreements, which lead to an endogenous internalisation of externalities (Krendelsberger 1996: 14). Since an explicit polluter cannot be identified easily, as discussed earlier, Pigou's approach can hardly be implemented for the upstream–downstream situation at rivers – who should pay the tax? Like Arthur Pigou, Ronald Coase also was concerned with externalities, in particular actions of companies that have harmful effects on others. Ronald Coase explains that economic analysts in such cases usually distinguish between a private and a social product of a factory (1960: 1). Economists have largely followed the argument of Arthur Pigou, who framed the issue of externalities in terms of blaming one party: an externality results if company A inflicts harm on company B. So, restrictions are proposed to restrain company A (Coase 1960: 1). But who is A and who is B in our case of an upstream and a downstream party? Ronald Coase avoids blaming one party as a polluter; rather he emphasises that:

> we are dealing with a problem of a reciprocal nature (1960: 1).

He regards externalities as a situation of two rival opportunities. How are the negotiations between the upstream and the downstream party framed? Game theory helps identify the framing.

The Flooding Game

In an upstream–downstream relation, the outcome of one actor depends on the actions of other actors. In particular, the downstream depends on the willingness and ability of the upstream to retain extreme floods. It is essential for the downstream to develop a strategy to respond to the upstream activities. Game theory helps to find and analyse such strategies (Cooter and Ulen 2004: 38). A game requires definitions of players, strategies of each player, and payoffs for each player for each strategy (Cooter and Ulen 2004: 39). For simplification of the following example, transaction costs are not regarded.

Rules of the game   The players in the flooding game are entities along a river. Such an entity can be institutional, like a municipality, an administrative district, region or even a state; such an entity can also be a land trust, an organisation or an assembly of landowners. Two essential preconditions are: first, that the players are empowered and able to decide upon the use of the land, and second that each player encompasses a cohesive area. For the following example, it is appropriate to imagine the players as two cities along a river. For simplification, only two parties are regarded in this game: Upstream and Downstream. Each party acts individually and for its own account. Resulting from the first part of this book, each party is in fact not an entity; rather each party consists of a social construction, one upstream, one downstream. But finally, each social construction produces a certain built environment: real constructions in floodplains (levees, houses, etc.). Hence, in the game, each social construction is hypothetically regarded as one player.

These built constructions in floodplains usually last for a long time. For that reason, the decision of what should be built is essential and binds the player for a long time. Players cannot easily change a decision. Thus, a wise strategy is needed. Each party has to decide between two opportunities regarding the use of their part of the floodplain, whereas it is assumed that both parties profit from housing projects in their own floodplain. Upstream can either build up housing projects or withdraw from building in flood-prone areas and instead provide retention volume for the sake of Downstream. The former opportunity implies that the flood wave will discharge faster from Upstream to Downstream due to strong and high levees, which protect the housing projects. Upstream chooses between the strategies 'retain' and 'accelerate' the flood. Downstream chooses between 'ignore' the flood risk and 'adapt' to flooding. Realising housing projects in floodplains disregarding the threat of inundation is the 'ignore' strategy (in fact, this strategy implies accelerating the flood to the next downstream through building high levees); risk adapted constructions of buildings and technical precautionary measures are the implications of the strategy 'adapt'. Downstream could withdraw from using the floodplains at all. This strategy would enable Upstream to do whatever he wants to do in the floodplains. Finally, each party decides between an individual rational strategy ('ignore' or 'accelerate') and a cooperative strategy ('retain' or 'adapt'), which considers the situation of the other party.

Each party receives payoffs from certain strategies. Four combinations of strategies are possible: 'retain/adapt', 'retain/ignore', 'accelerate/adapt' and 'accelerate/ignore'. Four conditions frame the payoffs.

The first condition is that 'retain' implies that Upstream acts for the sake of Downstream, considering what Downstream decides to do. Upstream provides as much retention volume as necessary to prevent damage from Downstream, but as little as possible in order to realise housing in the floodplains. Upstream has an interest in not wasting retention resources. Hence, if Downstream 'ignores', more retention volume is necessary than if Downstream 'adapts' (adapted housing project can absorb much more flood volume without damage). This is the cooperative strategy of Upstream.

The second condition implies that if Downstream decides to adapt, he concerns the behaviour of Upstream, and pursues, in this manner, the collective strategy. If Upstream retains, Downstream is able to realise expensive, but also lucrative, risk-adapted housing in the floodplains; if, on the other side, Upstream 'accelerates' the flood, and uses the floodplains for its own housing projects, Downstream merely can realise housing. Only few houses can be realised with high costs and a high risk of inundation.

Third condition: if both decide to profit most individually, the other party will not be taken into account. This means, if Downstream 'ignores' the flood, the housing area looks the same, regardless whether Upstream mitigates or accelerates the flood. However, Downstream must consider the probability 'p' of an extreme flood. A flood would reduce the profit of Downstream. If Upstream decides to 'accelerate' the flood, he does not regard the effects on the downstream.

Fourth: individual rational behaviour is able to produce the most individual gain (at least in the short-term). Upstream is able to realise housing projects in the whole floodplain if big embankments protect these areas. Downstream, on the other hand, gains the most if Upstream acts collectively rational despite Downstream acting individually rational. Then, cheap and extensive housing projects can be built with a relatively controllable risk of flooding. In the combination 'accelerate/ ignore', Upstream gains most, whereas Downstream receives less payoffs.

There is a remarkable asymmetry regarding the dependencies, since apparently the Upstream causes effects downstream and not vice versa. The situation of Upstream is (usually) not influenced by the decisions of Downstream (cases of tailback or specific bottleneck-situations will not yet be regarded); but the downstream situation significantly depends on the Upstream.

> Consequently, those who live upstream do not fear damages caused to them from downstream. In contrast, a fundamental reinforce of human norms is 'Someday you may be in my position, and how would you like if I … does not work (Scherer 1990: 20).

The strategy 'quid pro quo' – tit for tat – does not work, because of a lack of reciprocity (Scherer 1990: 20). This lack of reciprocity influences the game essentially.

Figure 2.7 shows an arithmetic example of a payoff matrix, considering the conditions above. The payoffs are a monetary gain (for example millions of Euros). Just for the simplification of the comparability, simple values are assumed, which represent the relations of the assumptions above. For the first condition, it is assumed that Upstream gains 7 for 'retain' if Downstream 'adapts', and 2 if Downstream decides to 'ignore'. The second condition leads to a payoff of 7 for Downstream in the case of 'adapt' downstream in combination with 'retain'

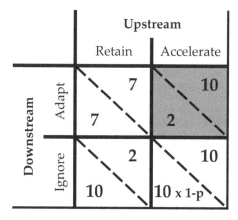

Figure 2.7   Flooding game payoffs

upstream; a payoff of 2 for Downstream in the case or 'adapt/accelerate'. Following the third condition, the gain upstream is 10, if Upstream 'accelerates' the flood wave, regardless whether the downstream 'adapts' or 'ignores'. On the contrary, for the reason of the mentioned asymmetry, a short-term rational decision downstream leads to 10 in the combination 'ignore/retain', but at least in the long run to a payoff of 0 in the combination 'ignore/accelerate' (because p becomes almost 1). The fourth condition explains why the particular maximum 10 for both can be achieved by short-term rational decisions, whereas only Downstream takes risk of losing payoffs. The following combinations of strategies can be played:

Accelerate/Adapt: for Downstream it would be most profitable if Upstream pursues a collective strategy. For Upstream, however, it is most tempting to act individually rational as well. Then, however, Downstream will have no payoffs in the long-term (probably it gains some short-term profits in the years between completion of the housing projects and the next flood). So, if Upstream indeed 'accelerates', Downstream should 'adapt' in order to achieve at least a profit of two. The economic welfare of the whole catchment achieves then a payoff of 12. Such situations can be observed in practice. The Netherlands discuss realising floating houses and even floating greenhouses and shopping malls. The Dutch try to gain as much as possible by a risk-adapted behaviour. Germany, France and Switzerland are the upstream parties. Indeed, retention takes place to some extent, but as a whole, these densely settled Upstreams accelerate flood waves and force the Dutch to adapt their housing projects (whereas it has to be admitted that the adaptive strategy of the Dutch is also owed to sea level rises, not only to river floods).

Accelerate/Ignore: if Upstream accelerates, the 'ignore' strategy pays off for Downstream if the probability of a flood is less than 0.8. In that case, namely, the payoff is $10 \times (1 - 0.8) = 2$, which is equal to the strategy 'adapt'. One could imagine that in reality this gain of 2 means that in the first years without a flood, Downstream could make a profit of 2 out of the housing project. If no flood would occur ($p = 0$), the full profit of 10 could be made out of the project. What does a probability of 0.8 mean? Flooding is a random event. The odds of occurring are independent of past occurrences (Cooley 2006: 105). Downstream, however, is not interested in the probability of an event in one year but rather in a period of y years (depending on the investment calculations for the housing project). If x represents the probability of a flood in a certain year. $1 - x$ is the chance that this event will not take place in a given year. The odds that an event will not occur in two successive years would be $(1 - x)(1 - x) = (1 - x)^2$. So, if $(1 - x)^y$ is less than 0.8, Downstream has an incentive to 'ignore'. For a centennial flood, this would mean: $(1 - 0.01)^y = 0.8$. Since $(1 - 0.01)^{22} = 0.8$, the probability that a flood does not occur within the next 22 years is less than 0.8; in other words, there is a 20 per cent chance of a centennial flood in the next 22 years. According to the statistics, that means that Downstream needs to plan housing projects so that they bring out a net profit of 2 within at least 22 years. But Downstream has to consider that, in the end, this is gambling with probabilities. So, in the short-term, the 'ignore' strategy is very attractive, in the long run the probability of flooding increases (precisely: the probability that a flood does not occur for a long period increases). The longer a project needs to be profitable, the higher is the risk of a flood. But the 'ignore' strategy is often applied in practice. Housing areas, industrial areas and further flood-sensitive land uses are often located downstream to other high-value uses, which are protected by levees, and thus accelerate the wave: 'urban waterfronts' along the rivers, financed by credit institutes, and promoted with slogans like 'Living near the River' are typical examples. In the long-term of our simple arithmetic example, the collective benefit of such allocations is zero.

Retain/Ignore: the combination 'retain/ignore' achieves the maximal gain for Downstream. However, this opportunity will dissatisfy Upstream, because it carries all the burdens and Downstream gets all benefit. This combination will only result in a situation with a very strong Downstream party, which has the opportunity to control or at least influence Upstream extraordinarily. Probably Upstream and Downstream are within the same administrative borders, and the decision power is with the downstream party. Another example for such a situation could be, if Rotterdam would stop all shipping upstream, in order to enforce certain retention volumes in Germany. Rotterdam's enormous important sea harbour, which delivers products from all over the world to the upstream hinterland and supplies an essential platform for trade, is an enormous bargaining power against Upstream – at least in this theoretical consideration, without regarding trade treaties and EU regulations. Within the arithmetic example, this combination reaches the second best collective gain, namely 12.

Retain/Adapt: the best solution from a collective rationality in this particular case is 'retain/adapt'. It achieves a common profit of 14. From the perspective of efficient allocation, as promoted earlier, this combination of strategies is preferable.

Finally, in a theoretical world without liability or other legal framings, economics predict the combination 'accelerate/adapt'; 'accelerate/ignore' emerges if short-term profits dominate decision-makers (the most probably strategies are highlighted in grey in the payoff matrix). In our example we had only two parties, when introducing many more, almost every party is an Upstream and a Downstream – so each has the incentive to 'accelerate' and 'ignore'. This is the described situation in the clumsy floodplains. The economically best result for the whole catchment is unlikely to happen.

Playing   Starting from the flooding game, we modify the rules of the game to see how redistributions of gains and losses may generate an economic more efficient allocation in the catchment area. We have already stated that it would be difficult to identify a polluter. Ronald Coase explains with an example of a cattle raiser and a farmer how liability for damage achieves an optimal allocation (Coase 1960: 5). Nonetheless, what would a liability yield? Assume an authority decides against the reckless Upstream who is affecting Downstream by accelerating the flood. From now, Upstream has to compensate Downstream for the losses. The distribution in

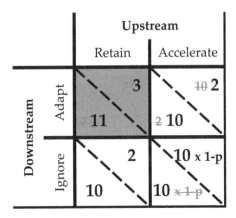

Figure 2.8   Introducing Upstream-liability rule

Figure 2.8 is the result (the right column changes). In the case 'accelerate/adapt', Downstream claims a payoff of 10. The remaining 2 are for Upstream. In the case of 'ignore/accelerate', the liability takes away the risk from Downstream. Suddenly, ignore or adapt are equal strategies. In each case, the probable losses are covered by the payments of the upstream. The risk is transferred to Upstream, who has now to estimate the risk. In long-term, 'retain' becomes the best alternative for Upstream in order to avoid payments to Downstream. The liability has another implication: the compensation of Downstream's losses through the liability rule deletes disadvantages of building in the floodplains. There is no economical reason for Downstream to reduce damage. The risk of flooding has no impact on allocation decisions, Downstream has an incentive to accumulate values, because Upstream takes the risk. Downstream has in this manner an incentive to waste resources, which is inefficient (Baumol and Oates 1988: 21).

Applied to the meat or crops situation, Ronald Coase describes in the article 'The problem of social cost', the cattle-raiser is like Upstream and the farmer matches the Downstream party. The cattle destroy the crop of the farmer – or, as in our case, the water masses destroy the houses on the floodplain. Ronald Coase introduced a liability of the cattle-raiser for the damage, which his cattle causes to the crop of the farmer. Now, the farmer cultivates more expensive crops on his field, so that the costs of the liability rise. Apparently, it is a misallocation to cultivate valuable crops in order to sacrifice it for compensation. In his article, Ronald Coase recognised this case:

> Suppose that the railway is liable for damage from fires caused by sparks from the engine. A farmer on lands adjoining the railway is then in the position that, if his crop is destroyed by fires caused by the railway, he will receive the market price from the railway; but if his crop is not damaged, he will receive the market

price by sale. It therefore becomes a matter of indifference to him whether his crop is damaged by fire or not (1960: 15).

But Upstream can offer a payment to Downstream for pursuing 'adapt' instead of 'ignore', and pursues himself the strategy 'retain'. Upstream could offer a payment of 4 of the original 7 to attract Downstream with the highest payoff in the matrix (11) for 'adapt'. Then, (almost) no damages happen and, in sum, the catchment yields a benefit of 14. The most efficient allocation is achieved.

Which possibilities does Downstream have without the liability rule? Ronald Coase shows that the allocation of resources will be the same in both cases: when the damaging business is liable for damage as well as the victim pays the firm for desisting imposing damage. The allocation depends on the benefit and the costs of damage. If the benefit is much bigger than the damage, in case of liability, the firm will accept these costs (because benefits still increase the costs), or in case the latter is not officially liable, then victims will not be able to pay off the firm (Coase 1960: 6–7).

The payoff matrix changes (Figure 2.9): Downstream would only pay Upstream for pursuing the 'retain' strategy. Upstream will only agree if it is at least not worst off with this option than with the other options. So Upstream agrees on every offer that assigns to it at least a payoff of 10. This implies that a payoff of 2 remains in the combination 'ignore/mitigate' for Downstream, the payment is about 8. In this situation, the combination 'ignore/mitigate' equals the combination 'adapt/accelerate'. What else could Downstream do? Downstream could offer to play the 'adapt' strategy and in addition offer a payoff of 3 for Upstream. It remains a benefit of 4 for Downstream in the case of 'retain/adapt'; Upstream receives a payoff of 10. Downstream could even heighten the offer up to 4, then Upstream has a real incentive to 'retain' and Downstream still has a profit of 3. In this case, the optimal allocation for the whole catchment is achieved without a central authority

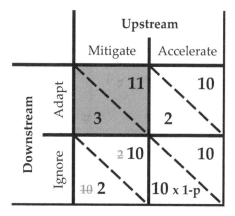

Figure 2.9   Downstream pays Upstream

or any liability rule. However, if Downstream considers only short-term profits and estimates 'p' lower than 0.8, Downstream will decide for 'ignore', Upstream will 'accelerate' then.

Finally, 'it is necessary to know whether the damaging-business is liable or not for damage caused since without the establishment of this initial delimitation of rights there can be no marked transactions to transfer and recombine them. But the ultimate result (which maximizes the value of production) is independent of the legal position if the pricing system is assumed to work without cost' (Coase 1960: 6–7).

Whether the conclusions of Ronald Coase are transferable to the Upstream–Downstream case, depend very much on the estimation of p. A sustainable treatment of the situation, however, regarding long-term effects, and in the long-term, p increases.

What happens, if some assumptions change? Until now, we assumed that the most benefit for the whole catchment could be achieved by 'retain/adapt'. Adapted housing, however, is expensive, in particular for existing structures; providing large retention areas might reduce profits more than primarily assumed. For that reason, we will now regard different payoffs: 'adapt/retain' yields only a payoff of 4 (or even less) for each party (see Figure 2.10). In short-term consideration, accelerate/ignore' becomes attractive, in the long-term, when p exceeds 0.8, the optimal allocation from the view of welfare are 'ignore/retain' or 'accelerate/ignore'. It does not matter whether Upstream realises large housing projects and the remaining strategy for Downstream is to 'adapt' ('ignore' implies lesser payoffs), or whether Downstream realises large housing projects, profits 10 and pays 8 of the 10 to Upstream for pursuing the 'retain' strategy. From the distributive point of view, finally Upstream profits 10 and Downstream receives a payoff of 2. Other measures than efficiency must be considered to decide on the allocation of the

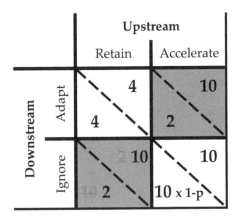

Figure 2.10   Expensive 'adapt/retain'

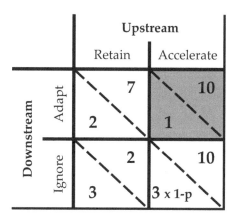

Figure 2.11   Valuable Upstream

housing project; however, the example shows that the parties will strive to the most efficient allocations.

The case will be different if Upstream and Downstream are not equal in their abilities to gain profit from housing projects. Imagine one party yields more land rent (because of better infrastructure, better marketing, better conditions for building, etc.). How will the parties distribute gains and losses, which allocation results?

Figure 2.11 shows a situation of an Upstream, which yields more benefit from housing projects than Downstream. Downstream yields a payoff of 3 maximum in the 'ignore strategy', reduced by the risk if Upstream plays 'accelerate'; the 'adapt' strategy is able to achieve 2 if Upstream retains, but only 1 if Upstream accelerates the discharge. In this situation, Downstream has no bargaining power to convince Upstream not to play 'accelerate'. Downstream remains the strategy 'adapt' to suffer no long-term loss. However, according to the Kaldor-Hicks efficiency, this is an optimal allocation, because the maximal benefit is achieved in the combination of 'accelerate/adapt'.

Vice versa, if Downstream yields more payoff from the housing projects, like in Figure 2.12. Upstream profits 3 maximum with the strategy 'accelerate'; 'retain' yields even less payoff. Downstream can realise housing areas and negotiate with Upstream about the costs for the 'retain' strategy. If Downstream builds adapted housing projects, it bids a payment of 3, but Downstream has an incentive to 'ignore', pay 2 of the payoff of 10 to Upstream in order to buy the 'retain' strategy. Downstream still has a profit of 8. The most efficient allocation, 'retain/ignore', emerges without any governmental intervention. Upstream has even a stronger position in this game, because in our example of two players, Upstream is a monopolist of retention areas. Upstream could claim for a payment of 6 instead of 2, because even then, a payoff of 4 remains for Downstream. At least in long-term consideration, this is still the best strategy.

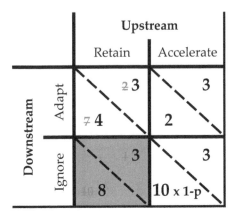

Figure 2.12   Valuable Downstream

In the real world, where more than one upstream party might provide retention areas, the payments would be a matter of negotiations. We can derive the general conclusion: either find a less valuable upstream, which you can convince by payments to retain floods, or offer a valuable downstream retention volume for an appropriate payment. In short: pay or swim!

Without central authority, even without liability, the power of the market steers the optimal allocation and distribution of flood protection. However, we cannot observe the described behaviour in practice. Probably, transaction costs for searching appropriate risk alliances of an upstream and a downstream party are too high. But even in obvious upstream–downstream relations, such agreements do not exist. Another explanation could be a low weight of *p*, then, the 'ignore' strategy is more attractive for Downstream. Another, very simple, explanation why the negotiations that Ronald Coase described do not take place is that the money for the payments is not available. The benefit of the upstream retention is a smaller loss in case of flooding. This benefit is difficult to measure because different individuals will perceive the benefit differently (Heiland 2002: 107–8).

However, these arguments cannot explain why an obviously more efficient technology, LATER, has not yet been implemented. In particular, because the basic idea of LATER is not fundamentally new. The Dutch calamity polders or the German *Notentlastung* have already been discussed, also upstream–downstream payments have already been a matter of research (e.g. Heiland 2002). What does it mean if a less efficient approach displaces an efficient floodplain management?

Technological Lock-in

Summarising the economics of LATER, flood protection is a scarce good, which produces benefits. LATER aims at an efficient allocation of these benefits. Following Kaldor-Hicks efficiency LATER requires a redistribution of the benefits, because the current allocation is already Pareto-efficient. Flood damage can be seen as an externality. However, internalising this externality requires an identifiable polluter and an unambiguous definition of victims. This, however, is almost impossible in the current situation. The flooding game shows how efficient allocations of flood protection can be achieved without necessarily identifying polluters or victims; rather mutual agreements on payments for certain strategies yield efficient allocations, whereas distributions might differ in accordance to the legal framing of the game (as the liability rule shows). Yet, the described negotiations on reallocation of flood protection do not happen in reality.

Competing technologies We are in a situation of two competing technologies: traditional levee-based flood protection and LATER. Both technologies aim at reducing flood damages. If two or more technologies serve the same purpose, which technology will be adopted and under which conditions will adopters choose which technology? Conventionally, economists answer this question with a long-run cost curve: the most efficient technology will be adopted because free market forces care for the adoption of the most efficient technology (Arthur 1989: 1, Thompson 2008a: 14). This has been described above by the flooding game.

Brian Arthur, professor for population studies and economics, was interested in the question, how an inferior technology might succeed other technologies (1989: 2). How can the less efficient levee-based flood protection persist against the more efficient LATER? 'Random events' (Arthur 1989: 1) 'can result us becoming locked in to a less efficient technology' (Thompson 2008a). In other words, free market forces do not necessarily lead to the best – meaning the most efficient – technology. Some randomness might support less efficient technologies. A situation can emerge in which users are entrenched in less efficient technologies, even though technologies that are more efficient are available.

The emergence of malaria, typhus and dysentery on the Rhine was such randomness. Johann Gottfried Tulla found that by building dams and by capturing the river, the diseases could be curbed. In addition, the wetlands became fertile farmland. From this point, the social construction began to work on the upper Rhine: floodplains became profitable, but due to the next flood, they became dangerous again, the Tullaian dams were heightened and improved – became levees – floodplains became controllable, then inconspicuous, and profitable again … The engine for the engineer-approach has been started then.

But why is it a lock-in? Michael Thompson explains technological lock-in with a very figurative example of the parallel invention of the water closet and the earth toilet. The consequences of the 'decision' for the water closet, is that:

> nearly 200 years later … the entire developed world is massively locked into the technologies of sewage and waste water treatment … The cost involved, now, in delivering the optimal perturbation, and getting us onto that much more

desirable (or should that be less crappy?) technological path, must run to trillions of dollars (Thompson 2008a).

What is the lock-in in our regarded situation? Flood protection by levees produces what the Dutch call 'paradox of security'. It means that levees pretend security, which justifies value accumulation behind them. Once they have been built, levees cannot be rebuilt (see Voigt 2005). The social construction of the floodplain sustains this effect. Finally, a levee-based flood protection is technologically locked-in. Due to time, this lock-in tightens, because more levees will have been built, more values will have to be protected.

Figure 2.13 shows the unit costs of a product, produced by different technologies. The more efficient a technology is the lower are the costs for one unit. For example, Technology A (levees) produces flood security. The costs of each unit result from building and maintaining the levees, but they contain also the costs of damage in case of failure. Increasing the benefit means reducing the unit costs. Technology B (LATER) produces flood security more efficiently, unit costs are lower. Traditionally, economists predict that technology B will then automatically by applied, if market forces work sufficiently. Instead, space for the rivers is shrinking. Most of the floodplains are occupied by intensive land uses. Once a levee is built, values of land behind it increase, farmers, residents,

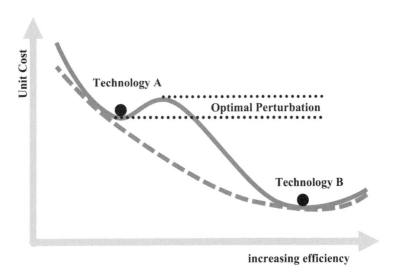

Figure 2.13   Technological lock-in

businesspeople, investors, all claim for using this land. Restoring floodplains for retention requires a lot of money to pay the current users for the restrictions the new land use will impose on the land. For several places, this is still possible, because floodplains are currently only used for relatively cheap farmland instead of high-end industrial use. But for more and more places, it will be quite expensive to introduce a retention-based flood protection. This is a technological lock-in. Such lock-in situations need interventions to push technology B. These interventions are called an optimal perturbation of the mechanism, which sustains the less efficient technology (Ellis and Thompson 1997: 214).

Why bother?   Probably, a perturbation is already at work. A paradigm shift is in progress: from flood protection to flood risk management, as discussed earlier. Water should no longer be excluded, rather it should be 'accommodated' (Wesselink 2007: 243). Floodplain management by LATER is a technological application of this paradigm shift. Paradigm shifts need time. Technological change requires a social and cultural process (Ellis and Thompson 1997: 209).

Changing patterns of activity in order to make space for rivers is a complex process (Moss and Monstadt 2008, Ellis and Thompson 1997: 213). Dik Roth and Jeroen Warner describe the struggle of Dutch water managers with integrative flood risk management. A 'shift from infrastructural to spatial flood protection measures' took place after the floods in the 1990s:

> While infrastructural works remain important, 'space' now became the keyword in Dutch flood policy. As people and water became competitors for limited space, 'water' is now finally also on the political agenda ... While earlier measures had only touched areas between dikes, this time Rijkswaterstaat was going to intervene behind dikes. Spatial planning landed the department in an unknown arena where project design does not take place in a top-down fashion but many interests are involved in a consensual process of negotiation (2007: 520–21).

Anna Wesselink describes the Dutch flood defence system as politically locked-in, where 'it is unable to escape its control paradigm for flood risks' (Wesselink 2007: 246). The vulnerability of the Netherlands, she concludes, is through a 'combination of managerial and political decisions ... socially constructed' (Wesselink 2007: 246).

So, the paradigm does not occur yet with high pressure, but it occurs. So, why bother? Although contemporary floodplain management is clumsy, it is viable; and probably, we do not need a more efficient technology today. Currently, extreme floods are relatively rare. Probably, society will change its patterns of activity when the problems with these patterns exceed the advantages.

Clumsiness – sustaining unsustainability   Clumsiness yields advantages! Clumsy floodplains provide a high level of deliberative quality (Thompson 2008a: 13).

> Clumsiness concerns both the effectiveness of attempts to tackle major social
> problems and the legitimacy of this process. (Verweij et al. 2006a: 20)

It encompasses varying perceptions of the world. Each rationality is heard, and
to each it is responded. This improves both the democratic quality and the policy
outcomes. Clumsy solutions combine two advantages: they are widely accepted
(Ney 2007: 322–3), and they are effective in achieving the democratically chosen
goals, which means they contribute to 'the alleviation of pressing, practical
collective problems' (Verweij et al. 2006a: 20).

The pressing, practical collective problem on which all rationalities agree on
is that floodplains must be managed in some way. But which way is a matter
of discourse and disagreement. Nonetheless, stakeholders in riparian landscapes
found a way to manage floodplains with all its different forms of appearance:
water managers build strong levees; landowners use floodplains profitably; when
a disaster occurs, society commonly defends the water masses and voluntary
donations and governmental aid supports reconstruction. As long as extreme
floods are rare, current clumsy floodplain management works well. Stakeholders
complained not fundamentally in the interviews about the current floodplain
management. Contemporary, the benefit of flood protection with levees is
estimated higher than the costs it produces (namely damages) (Strobl and Zunic
2006: 384). The pressing, practical collective problem is alleviated for all involved
stakeholders to an acceptable degree. So, as long as each party perceives clumsy
floodplains more convenient than inconvenient, let it be. But for how long?

An important aspect of land management is sustainability (Needham 2006: 139).
Clumsy floodplain management is not sustainable because it depletes retention
resources. LATER, however, provides technological flexibility through the mixture
of hard engineering levees, adaptive measures and economic instruments.

The social construction of floodplains works like a risk-producing machine: it
yields levees along the rivers, which yield further urban development in floodplains,
which need additional levees, which yield additional urban development … A
machine works as long as enough fuel is available. The fuel of this machine is the
land in the floodplains, which is limited. The machine drivers are the rationalities
within the social construction. But when – almost – each hectare of land in the
floodplains is used for residential, commercial and industrial purposes, no space
for the rivers is left and rivers are captured by very high levees. Maintaining these
levees will cost an enormous amount of money and resources; uncertainty remains
according to the effects of climate change and crevasses.

Already today, engineers emphasise the necessity of strong and high levees:

> As a Dutch civil engineer put it recently: 'there is one big lesson for the
> Netherlands from [Hurricane] Katrina: we should never allow this to happen. If
> you want to live in these areas, you have to protect them'. He should know that
> the word never does not exist in risk management (Wesselink 2007: 242).

Indeed, the engineer referred to coastal floods, but this statement fits as well to river floods. But once the retention volume in the floodplains is depleted, it will be very difficult to restore it. Already today, it is a very difficult task in catchments like the Rhine, the Danube or the Elbe really to restore floodplains. Levee-based flood protection results in an:

> inescapable technological lock-in, where the only choice available is to continue with ever-growing dikes and dams (Wesselink 2007: 243).

The social construction counteracts the paradigm shift. The earlier an optimal perturbation takes place, the easier will an optimal perturbation cope with the lock-in, because due to time, the social construction increases the necessary perturbation (less space for the rivers is left for retention). An immediate intervention is required, whereas this intervention should minimise technological inflexibility (Ellis and Thompson 1997: 209). What would an optimal perturbation of the technological lock-in look like?

Chapter 3
Responsive Land Policy for LATER

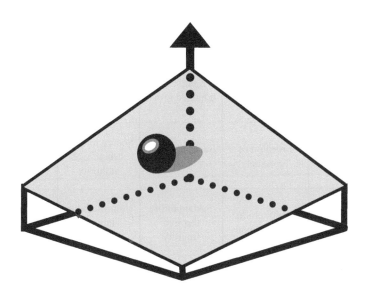

Polyrational Land Policies for LATER

Large Areas for Temporary Emergency Retention – LATER – can efficiently cope with extreme floods. The social construction, however, entrenches a technological lock-in into a levee-based flood protection paradigm. Surmounting this lock-in requires a change of the land uses in floodplains. This requires interventions by land policy in private properties. Each municipal or governmental intervention in value, use or distribution of land can be denoted as a land policy (Davy 2005: 117). Attempts of intervening in land use should not only achieve efficiency but also be effective in democratically chosen goals (Needham 2006: 139). A land policy for LATER must effectively achieve an optimal perturbation of the technological lock-in.

An intervention in clumsy floodplains can only be effective if the four stakeholders support the intervention. However, spatial planning and interventions in land uses bear crucial conflicts (Davy 1999: 103). The stakeholders will only support interventions that they perceive at least as rational. Four rationalities determine the social construction and influence the actions of the stakeholders in different phases. The rationalities are biased. Michel Schwarz and Michael Thompson postulate that:

	Individualists 	Egalitarians 	Hierarchists 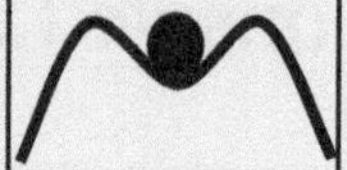	Fatalists 
Social context	Negative group/ negative grid	Positive group/ negative grid	Positive group/ positive grid	Negative Group/ positive grid
Way of organising	Ego-focused network	Egalitarian bounded group	Nested bounded group	Margins of organised patterns
Latent strategy	Freedom of contract	Survival of the collectivity	Secure internal structure of authority	Survival of the individual
Systems properties	Exploitability through fluidity	Sustainability through fragility	Controllability though orderliness	Copability through chaos
Institutions	Deserve individual's profit	Affirm shared opposition to outside world	Warrant correct procedures	-
Salient risks	Market failures	Irreversible catastrophes	Loss of control	-
Learning style	Trial and error	Trial without error	Anticipation	Luck
Ideal of fairness	Equality of opportunity	Equality of result	Equality before the law	Not on this earth
Model of consent	Implicit consent	Direct consent	Hypothetical consent	Non consent

Figure 3.1   Characteristics of the rationalities

our knowing is biased, our acting is biased, our justifying is biased. Bias is to organizing as gravity is to walking about: we would be in a bad way without it. So bad, in fact, that we would have no society and no culture: no social relations and no cognition. Each of the four political cultures is a consistent package of biases (1990: 61).

Each rationality – the hierarchic, the individualistic, the egalitarian and the fatalistic – would pursue another land policy, for LATER. Clumsy enough, each land policy will be rational, following its own distinct rationality.

Finally, many (poly)rational land policies emerge. Each rationality has its own idea how property-focused risk alliances for LATER should be initialised and governed. Four proposals for monorational land policies will be presented in this chapter, one distinct land policy for each rationality. Stereotypes of the rationalities are summarised in the book *Divided We Stand* (Schwarz and Thompson 1990). Other literature about Cultural Theory serves as a guideline for developing the land policies. Figure 3.1 shows some characteristics of the rationalities, borrowed from Michel Schwarz and Michael Thompson (1990: 66–7).

According to the hierarchic rationality, a constitutional land policy will be presented, owing to the strong grid and group dimension of this rationality. Individualists pursue a competitive approach, following their characteristics. The land policy of egalitarians is cooperative. Fatalists are composed, in accordance their land policy is composed as well. In this manner, a certain attribute can be assigned to each land policy: the competitive, the cooperative, the constitutional and the composed.

Each land policy is derived from a prototype. This prototype is borrowed from different disciplines (economics, planning, etc.). After the description of the prototypes, each proposal for a land policy according to the respective rationality is described schematically. The implementation is described in three steps. Finally, land policy implications are going to be described, before the four land policies are compared with each other.

Competitive: Privatised Flood Protection

In discussions about LATER with practitioners in the *Hochwassernotgemeinschaft Rhein e.V.*, it became obvious how municipal administrations are entrenched in their belief that flood protection is a governmental task. The competitive land policy questions this belief. LATER aims at enhancing efficiency. Efficiency is a domain of the market, not the government. Why should we not use market forces to achieve efficient allocations of LATER?

> There is an economic argument that free trade (in rights) leads to maximum efficiency in the use of scarce resources. This argument, in its most naive form, leads to the political conclusion: there should be no interference in free trade (Needham 2006: 11).

How the market leads to efficient allocations can be explained with the example of a lake and the fishermen by William J. Baumol and Wallace E. Oates.

The lake and the fishermen For a market, well-defined property rights are essential (Needham 2006: 24, 32). William J. Baumol and Wallace E. Oates state:

> The source of an externality is typically to be found in the absence of fully defined property rights. And this implies that in some instances the distortions resulting from an externality can be eliminated through an appropriate redefinition of such rights of ownership (1988: 28).

William J. Baumol and Wallace E. Oates explain:

> Consider a lake to which all fishermen have free access (Baumol and Oates 1988: 27).

For LATER, the lake is the catchment area of a river. The fishermen are the landowners:

> The haul of one fisherman reduces the expected catch of others, so a detrimental externality is present (Baumol and Oates 1988: 27).

The haul of the fishermen corresponds with profitable land uses in the floodplain: the more values are accumulated in floodplains (potential damage), the higher and stronger levees are needed and the less space for the rivers is available for the retention of extreme floods. Using floodplains for urban development reduces the benefit of the land of the others (because if they accumulate values as well, they might suffer a loss in the next flood):

> Suppose that, instead of introducing an entry fee, the lake were transferred from public to private ownership, perhaps through some sort of auction. Suppose, moreover, that the new owner seeks to maximize profits from fishing activities on the newly acquired lake (Baumol and Oates 1988: 27).

The private owner will hire fishermen. He will pay wages in return of the haul. Translated to floodplains, one owner of the lake is the owner of the floodplains. He grants tenures for using the land. The owner of the floodplains pays the tenants in return for the land rent. The lake-owner will engage fishermen up 'to the point at which the value of the marginal product equals the wage' he is able to pay (Baumol and Oates 1988: 27):

> Thus, the private ownership outcome will be socially efficient (Baumol and Oates 1988: 27).

For LATER, it is not necessary to sell the whole floodplain to one owner. LATER needs efficient risk alliances for reducing the damage through extreme flood events. To explain it with the lake model, it would be adequate to sell only the fish in the lake to a private stakeholder. The stakeholder owns the fish in the lake. To profit from the fish the owner must hire fishermen. For LATER: the fish are (a part of) the land rent. The stakeholder owns this part of the land rent. The wage landowners receive is flood protection against extreme floods (which is avoided loss). Assume, in the lake case, there are successful fishermen and less successful fishermen. In order to maximise benefit, the lake-owner must hire the successful fisherman and pay them well, and he must fire the less successful fishermen or pay them less. This means: successful landowners, who produce more land rent (which partly belongs to the stakeholder), receive more protection against extreme floods. Protecting less successful landowners would be a waste of money (or protection) for the floodplain stakeholder.

Finally, the owner will bring floodplains into a situation where valuable land uses are highly protected, whereas less profitable land uses will be less protected. This is the desired situation according to the allocation of the scarce good flood protection. In final consideration, the floodplain-owner (or the owner of such rights in a part of the land rent) will figure out that LATER achieves an efficient allocation of flood protection according to extreme floods. The lake and the fishermen become a prototype for a land policy for LATER, based on private ownership. In this manner, privatisation of flood protection achieves efficiency.

Privatised flood protection   Deregulation and privatisation of network-bound public services has been a trend in contemporary politics since the 1970s (Scheele 2006): in Germany, the national post became *Deutsche Post AG*, *Deutsche Bundesbahn* became *Deutsche Bahn AG*, the communication sector is privatised, the energy sector becomes increasingly deregulated, for some years, private enterprises have been capturing drinking water supply and sewage systems. In the late 1980s, privatisation was en vogue in Western Europe (Segeren et al. 2007: 13). Like flood protection, the affected sectors had been traditionally public tasks before. Competition was not welcome in these sectors. Public engagement allowed fast and nationwide provision of the services (Scheele 2006). So, public flood protection yields nationwide levees. Today, virtually every river bigger than a runlet is captured by levees, each landowner has access to a minimum of flood protection. This, as described earlier, will lead to a technological lock-in.

The reasons for privatisation have been problems in financing the services with public money, claims for economic efficiency, and the use of new technologies (Scheele 2006). LATER aims at enhancing efficiency and disburdening public households. LATER serves private interests (which is indeed a paradigm shift from flood protection as a public good to flood protection as private service) and demands the use of new technologies. The reasons to privatise network-bounded public services are similar to the demands of LATER. However, privatisation is not unproblematic in accordance to nationwide supply guarantee. This aspect is discussed particularly in the energy sector (Scheele 2006), but also in other sectors. Rural areas might be less profitable for providing network-bounded infrastructure such as high-voltage power lines or high-speed railway connections. This is in fact not a counter-argument against privatising flood protection for LATER; rather it is a pro-argument. LATER even aims at disturbing nationwide supply of flood protection by levees for everyone, but rather only the most valuable uses should be protected by levees. Finally, strong arguments support privatising flood protection.

Implementation   An individualistic land policy for LATER could be accomplished in three steps:

First, the state offers concessions for flood protection: changing responsibility for flood protection requires action of water management agencies, which are currently responsible. According to the lake and the fishermen example discussed above, the future stakeholder must be responsible for protection measures, because this corresponds with the wages of the fishermen. In addition, the future stakeholder must be empowered to collect some kind of fee for providing flood protection. This fee corresponds with the haul of fish. It is an inherent threat of privatisation that they increase economic welfare at the expense of sustainability or social justice. For that reason, privatisation should be accompanied by regulations (Segeren et al. 2007: 20). So, for avoiding a lock-in into a monopolist market, which can hardly be regulated, it would be good not to sell flood protection to a private instance, but grant concessions, which last for a certain period. After this period, competition readjusts probable undesired developments of the market.

This period, however, must be very long in order to heighten the probability of an extreme flood within this period. A period of at least 50 or even 100 years seems suitable. It depends on the risk prognoses for a catchment. The probability of an extreme flood must be at least high enough to set incentives to initiate preparations for such events. In addition, the concession area must be large enough to realise LATER, the best would be catchment-wide concessions. These concessions empower the concession-holder collecting fees from individual landowners within the catchment area. The concession-holder in return is responsible for the flood protection in the whole catchment area, and liable for damages through floods (including extreme floods). To prevent insolvency of concession-holders in case of flooding, the concession-holder has to put a catastrophe deposit aside to a public trustee, large enough to cover the damages of the next flood event.

In the second step, concession-holders strive for an increase of efficiency: the award of the concession takes place after all potential concession-holders staged their bid, and approved solvency for extreme floods. The bids of different concession-holders distinguish in the fees, which they want to collect from landowners. The calculation of the fees is crucial. Gains and losses of the concession-holder depend on this calculation. Between bidders for the concession, a competition about the fees emerges. If fees are appraised too low, concession-holders make less profit. If fees are calculated too high, another concession-holder underbids them and gets the concession. Measuring the fees includes the costs of an extreme flood. In such cases, concession-holders must balance which areas will suffer less damage from an inundation. These less costly areas will become LATER.

Third, concession-holders negotiate about payments: in order to reduce costs, the stakeholder must negotiate about LATER with landowners. A negotiated agreement about a controlled flooding of a certain area will save money. Landowners and concession-holders negotiate then about compensations and reductions of fees of landowners within the flooding areas. In addition, concession-holders must estimate higher fees for a better protection of landowners downstream compared to the landowners of the retention land. Finally, the distribution of the gains and losses through LATER results from negotiations between landowners and concession-holders.

In final consequence, every landowner has to 'pay or swim'. The 'invisible hand of the market' steers the allocation and the distribution.

Land policy implications   A market is a system of voluntary exchanges of goods and services (Segeren et al. 2007: 12). It yields efficiency if five conditions are met: private property, freedom of contract, free access to the market for all, currency, competition (Hartkopf and Bohne 1983: 78–82). The state has to optimise these conditions because:

> market exists and works thanks to rules created by the law maker and enforced
> by state agencies (Needham 2006: 24).

The five conditions are met for the proposed competitive land policy as follows:

Markets can only trade rights, which are recognised by a court (Bromley 1991: 35). Inundation easements are the traded rights. The concession-holder is the beneficiary, and the landowners within LATER are the suppliers of the easement. The trade needs to be registered in the land cadastre, so that the right is legally enforceable. The right adheres then to the real estate. The protected downstream landowners receive a contractual guaranteed liability in case of flooding in return for the fees they pay to the concession-holder.

The state must guarantee freedom of contract. Each landowner and the concession-holder must be able to negotiate about LATER. However, the state regulates the market to ensure that no humans will be endangered through the contractual agreements between the landowners and the concession-holder. The contract has to consider appropriate forewarning for evacuation of the retention areas before flooding. In addition, the state has to determine suitable contract penalties if the concession holder does not fulfil the concession – e.g. if he will not maintain the levees or claim for higher fees than negotiated. The state has on the one hand the duty to monitor the contractually negotiated agreements; on the other hand, the state has to support the concession-holder for instance as a mediator in difficult negotiations with landowners. The latter role of the state is important from the perspective that a private company cannot enforce agreements by governmental instruments. The task to monitor the concession-holders' duties could mean that the concession must be renewed in certain periods – e.g. every 10 years. Then, the renewal process could give other competitors the opportunity to displace the former concession-holder, if they have not fulfilled their duties or another competitor makes a significant better offer.

Free market access is important for the competition about the concessions. The state has to ensure that potential concession-holders are able to cope with the task to provide protection, collect fees and pay in case of flooding for the damage. Only big companies meet these conditions. However, each company, which matches these conditions, may give a bid.

Currency is the condition that is fulfilled in the most countries – but the more stable the currency is, the better is the situation for efficient solutions, because the concessions are for very long periods. Thus, not only the bare existence of currency is an issue, but also its stability.

The state must also ensure fair competition between bidders for the concession. Each bidder must have the same chance, which is very in line with the individualistic rationality (equality of opportunity). The role of the state is crucial even in this land policy of individualists, who are in favour of a less structural grid; without such interventions a 'naive form' (Needham 2006: 11) of free trade remains.

Privatising flood protection with the proposed concessions is an individualistic land policy. It emphasises a weak structural grid with a negative group adhesion. Indeed, tendering and controlling the concession requires hierarchy, but the institutions serve for an organised competition. As depicted in Figure 3.1, this matches the individualistic commitment to institutions. Granting concessions to private enterprises for a period reflects the 'trial and error' approach of individualists. For the land policy, the functioning of the market is essential. Freedom of contract and equality of opportunity characterise this approach. The competitive land policy is an individualistic way of managing floodplains for LATER.

Cooperative: LATER Land Trusts

The egalitarian rationality emphasises cooperation. Egalitarians act in a market in homogeneous groups, competing with other groups. The well-being of the community is important for individual decision-making in the egalitarian rationality. 'All for one, one for all' – the egalitarian land policy emphasises the community. It sustains private associations with a low structural grid but a high responsibility for the group. Land trusts for land conservation in the United States are an example of how such an association can work. They are cooperative and informal institutions of this land policy.

Land trusts for land conservation   For the preservation of open space, special voluntary land trusts exist in the United States. The Land Trust Alliance is an umbrella organisation for these organisations. They describe a land trust as a:

non-profit organization that, at all or part of its mission, actively works to conserve land by undertaking or assisting in land or conservation easement acquisition, or by its stewardship of such land or easements (Land Trust Alliance 2007).

Land trusts are described as independent and entrepreneurial. They work with landowners who are interested in preserving open space. The US land trusts acquire easements or sometimes they acquire full property in open space. Since they have a non-profit tax status, land trusts can offer landowners a variety of tax benefits. Landowners sometimes donate land or easements to the trusts to save taxes:

> Donations of land, conservation easements, or money may qualify you for income or gift tax savings (Land Trust Alliance 2007).

Land trusts are independent from municipalities and governments. They are not embedded in a strong structural grid:

> Land trusts often work cooperatively with government agencies by acquiring or managing land, researching open space needs and priorities, or assisting in the development of open space plans. Land trusts are very closely tied to the communities in which they operate (Land Trust Alliance 2007).

So, the American land conservation egalitarians ally with the hierarchic rationality, but:

> because they are private organizations, land trusts can be more flexible and creative than public agencies – and can act more quickly – in saving land (Land Trust Alliance 2007).

Could these trusts be a prototype for an egalitarian land policy for LATER? Private land trusts for LATER would need municipal support as well. Public agencies have data about the hydraulics and other relevant information (e.g. in the cadastral register). In extension to the American prototype, land trusts for LATER do not only conserve vacant land but also regain already lost retention areas. If the public agencies deliver relevant information and help to establish independent and private land trusts, this could be a powerful tool to realise LATER. They are not bound to administrative proceedings or political demands, land trusts could profit from other spatial developments, for instance shrinking cities: shrinking cities contain vacant land, which is low in value and difficult to sell (Davy 2006), land trusts could offer tax savings for landowners, which donate inundation-easements.

An essential difference between the land trusts for land conservation in the United States and the LATER land trusts is the motivation of the trusts. Land conservation land trusts act 'to benefit communities and natural systems' (Land Trust Alliance 2007). LATER land trusts are driven by self-interest of landowners who want to save money (downstream) and gain money (upstream), respectively, in the next extreme flood event. This difference is relevant, because the land conservation land trusts in the United States pursue a non-profit issue. Landowners will probably donate land voluntarily for the common purpose of land conservation but they will not donate their land for LATER, because it creates advantages only to certain landowners. Another essential difference is the need for coherent land. Plots for land conservation may be located in different areas, but land trusts for LATER requires particular and coherent areas. The trusts have to meet the requirements.

LATER land trusts Civil initiatives are already important stakeholders in floodplains. Landowners form initiatives to express their demands for better flood protection commonly. Some initiatives last a long time, the initiative in Cologne-Rodenkirchen for example is very active and present in the media. Typically, such initiatives are launched after a flood, when the egalitarian rationality dominates floodplains and landowners perceive floodplains as dangerous terrain (see Figure 1.3). Civil initiatives are most often quite well informed and open-minded for new approaches to flood risk management. The civil initiative in Cologne-Rodenkirchen, for instance, is fascinated by the idea of LATER, as it became apparent in a discussion with the chair of the initiative.

LATER land trusts could build on such initiatives of organised landowners. Landowners are evidently affected most by an extreme flood. They suffer losses and they know how big the damages are. Why should the state or anyone else care for them? The state is not legally liable for damages from natural hazards. Landowners should and can take responsibility for the risk they produce and suffer. Landowners do not stand alone with this responsibility; they can establish land trusts and care for large retention areas upstream for extreme floods. How can such land trusts become a viable element of land policy?

Implementation In the first step of implementing LATER land trusts water management agencies announce that public flood protection cannot cope (any longer) with extreme floods. It has to be clear that the state cannot afford, for example, 20-metre-high levees in order to protect a few single-family houses against extreme floods, and that extreme floods will occur more often and more intensely in the future. Moreover, in the face of extreme floods, landowners must be aware that the state and the society might help in disasters like in 2002 at the River Elbe, if such events occur very rarely, but frequent extreme floods imply less governmental and voluntary support. Abolishing governmental compensation can indeed stimulate risk-adapted behaviour of landowners, as an empirical study for the Netherlands points out (Botzen et al. 2009: 27). The first step is to initiate land trusts for LATER. Therefore, a high risk awareness is necessary, so that private landowners know the risk of extreme inundations. In dangerous floodplains, it would be easier to convince landowners to contribute to such land trusts. This risk awareness and the willingness to alleviate the risk is the starting point for the voluntary establishment of land trusts.

After that first step, landowners will establish land trusts: the knowledge about the potential damage through an extreme flood and the increasing threat of such floods produce an interest of landowners to protect against extreme floods. Municipalities could encourage and support landowners in establishing land trusts. However, the establishment of the land trusts neither lies in the competence of the municipalities, nor of other public agencies; in fact, landowners are responsible on their own.

The land trusts have to challenge a difficult task: LATER requires a coherent area; to protect a particular area downstream, a very large coherent area in the right location upstream is required. Therefore, two tasks have to be carried out. Namely, every landowner within the potential retention area has to be convinced to sell inundation easements, and enough landowners have to be found downstream to pay for additional retention upstream. In other words, cooperation and conviction of landowners is essential for LATER land trusts.

On the other hand, single landowners downstream will not be able to purchase inundation easements in an appropriate size upstream, because LATER contains land of many landowners. The bigger the community of solvent downstream landowners, the more retention is going to be available. On the other hand, the more landowners are involved, the more complicated the negotiations will be. In this situation, land trusts could be a vehicle to facilitate the negotiations, if they are empowered by the landowners to negotiate about LATER with upstream parties.

The upstream landowners can only sell inundation rights to the downstream land trusts in certain coherent sizes. It is helpful if upstream landowners, who earn money from selling inundation easements, establish land trusts as well for the negotiations with the downstream land trust. This matches the egalitarian rationality, acting in markets as corporate groups.

Third, land trusts negotiate about payments: a landowner can either sell or purchase inundation rights. So, free riders upstream are worst off in this system: landowners who would not like to sell inundation rights suffer a lack of flood protection without any compensation – because the state will not protect them against extreme floods. Obstructers downstream actively weaken their own position by not participating, because less LATER can be made available without their contribution. In the egalitarian scheme, collectivity is essential for its implementation.

In results, risk-alliances in a catchment area are only possible to realise LATER if efficiency of flood protection in the catchment is enhanced.

Land policy implications Land policy by land trusts is based on voluntary cooperation of landowners. This land policy depends on the ability of a community of landowners to establish viable trusts. Two types of land trusts are necessary: 'sellers' sell inundation rights, they supply LATER; 'buyers' purchase inundation rights upstream, they consume LATER. Potential members of both types of trusts may be located in the same neighbourhood. In the same neighbourhood, land trusts for selling LATER and land trusts for consuming LATER could compete with each other. Despite this competition and individualistic attitude, the form of this competition is not individualistic, but rather egalitarian. Bounded groups, namely land trusts, compete, not individuals.

Land trusts have to cope with both: internal competition between different landowners about selling and buying inundation easements within their area, and achieving mutual agreements with land trusts both downstream and upstream about such rights. How can collectives of landowners achieve such agreements without a central or even coordinating institution? Land trusts advocate for the landowners, searching for retention areas upstream or selling inundation rights for downstream. If 'sellers' want to offer their area for retention, they must collect enough money from the downstream 'buyers' to cover their own costs as well as paying out the potential 'buyers' in their area (which want to sell retention easements upstream). In an area, competition between 'sellers' and 'buyers' emerges. The result of this competition decides if an area becomes LATER or whether it will be protected by LATER upstream.

In a valuable area, like for instance in Cologne, landowners who want to sell inundation rights to Düsseldorf (downstream to Cologne) could not collect enough money to pay off the citizens of Cologne for the damage in case of retention, because Cologne is not an efficient allocation for LATER. On the contrary, the landowners of valuable urban sites in Cologne are able to establish a powerful land trust, which gathers enough money to pay for a large retention area for extreme floods in the rural *Siegaue* (upstream to Cologne). Finally, the cooperative scheme for LATER needs to establish robust land trusts. These communities must then cooperate in the market.

Land trusts must consider free riders. If, for example, a land trust in Cologne has achieved an agreement with a land trust in the *Siegaue* about LATER, the land trust of landowners who wanted to sell inundation rights to Düsseldorf will not participate in the payments. They are free riders. As long as enough landowners pay for LATER, free riders are acceptable. However, free riders are a moral problem. Fewer landowners will be willing to pay for the free riders. If they and payers are almost equal in number, the problem of inundations is not pressing enough; the area is not valuable enough to protect it by LATER. Free riders are in this manner not a problem; rather they are an indicator for efficiency.

For obstructers, the case is similar. An obstructor is a landowner in a potential location for LATER, who neither wants to sell inundation rights, nor wants to join a land trust, which buys LATER upstream. Land trusts lack compulsory instruments like expropriation. Besides, expropriation is only allowed for public purposes (article 14 III GG), but land trusts act self-interested. Only moral and economic arguments remain for land trusts.

Economic arguments are quite strong. Since the state is not liable for damages through an extreme flood. So, landowners are not entitled to claim for compensation or protection against such events. Without LATER, it is difficult to predict which properties will suffer damage in the next extreme flood. LATER makes this damage predictable. If the landowner remains obstructing despite appropriate offers from downstream, and despite all neighbours being willing to sell inundation easements (because they cannot afford to buy LATER for themselves upstream), no deal will be possible. The neighbours are not entitled to reduce flood protection (LATER requires a construction for a controlled flooding, which actually reduces the protection level). As discussed earlier, landowners may sue if the existing protection level is reduced without their agreement. If no such agreement exists, the risk of being inundated remains, but in case of flooding, no compensation will be paid. The obstructor is better off to agree, join the 'supplier' land trust, in order to receive compensation in case of extreme floods. Landowners within a hydraulically cohesive area are all virtually in the same boat. Hence, they are better off to cooperate.

This stress on cooperation makes this land policy an egalitarian land policy. The here described egalitarian land policy mirrors the characteristics of the egalitarians (see Figure 3.1). Land trusts are organisations with a weak structural grid – hierarchic instruments are not available – and a strong group adhesion. Land trusts need a collective behaviour to survive in a dangerous world. Egalitarians believe that although markets often fail in achieving efficiency, governmental interventions cannot guarantee the achievement of efficiency either (Buitelaar 2003: 316). For them, the network is the preferred way of organising. So, cooperation through direct consent is most important to cope with extreme floods.

Constitutional: Mandatory Protection Readjustment

> Assembling land to supply the variety of public needs is a problem shared by local government, planners, and developers across the world (Alterman 2007: 57).

Assembling large retention areas for flood protection affects property, law and economics of land. A tool that can operate across these issues is land readjustment (Alterman 2007: 57). It reallocates and redistributes rights in land. LATER reallocates protection and redistributes gains and losses of protection. Land readjustment can therefore be adapted to the LATER requirements.

> Mandatory land readjustment helps put land use plans into practice (Davy 2007: 38).

An alternative hierarchic planning instrument to provide land for public needs is expropriation: expropriation, however, may not be used for every purpose, and requires compensation and therefore is expensive for public households (Alterman 2007: 71–2). Due to the size of LATER, expropriation would be too expensive, and it might be questioned whether an expropriation for the purpose of scarce inundations is a proportional instrument. Regulatory takings, downsizing and compensation rights are hierarchic tools for assembling land, but like expropriation, the required compensations burden the public. On the contrary, 'land readjustment can help preclude compensation claims' (Alterman 2007: 75). It partially internalises increases and decreases in property values (Alterman 2007: 75). Finally, land readjustment is an instrument for reallocation and redistribution of gains and losses, without imposing on public households (Needham 2007b: 116).

Readjustment is an appropriate prototype for a hierarchic land policy. In Germany, this mandatory land readjustment – *amtliche Umlegung* – is legally founded in the Federal Building Code (BauGB). § 45–79 BauGB contain detailed instructions for readjustment proceedings. Mandatory land readjustment is a formal instrument to put plans into practice. It is part of the nested German planning system. It controls estates through inherent orderliness, based on correct procedures. Each landowner is equal before the law. German mandatory land readjustment fits to the hierarchic rationality, regarding the characteristics in Figure 3.1.

German Mandatory Land Readjustment

> Land readjustment is an independent, free-standing instrument with its own internal rationale (Alterman 2007: 80).

What is this rationale in German mandatory land readjustment? Land readjustment is a procedure to create suitable building plots for the designated land use according to shape and size. In the past, readjustments were often used to rearrange a built area in situations where the buildings were destroyed by disasters; in particular after the Second World War land readjustment helped to rearrange the sites for new requirements (Dieterich 2000: 47). Today, land readjustment is used for reshaping agricultural sites (*Flurbereinigung*) or developing urban areas (*Baulandumlegung*). Both types of readjustment require cooperation of landowners in many countries, for example in the Netherlands (Needham 2007b: 117). In Germany, however, mandatory land readjustment may be compulsory; it empowers planners to enforce binding land use plans if landowners are not willing or able to modify their boundaries voluntarily (Davy 2007: 39). This makes mandatory land readjustment a powerful hierarchic tool for plan implementation in Germany. But in final consideration, the rationale of land readjustment aims at producing a net gain for as many of the involved landowners as possible (Needham 2007b: 118) – no matter whether this 'happiness' (Davy 2007) is mandatory or voluntary – through a more suitable arrangement of sites within the land readjustment area. 'More suitable' can be translated into 'more efficient', owing to the economic focus of land readjustment: usually, the fair market value is the measure for proportional redistribution, except the land is almost equal in value in the whole readjustment area (§ § 56–8 BauGB).

A 'pooling agent' (Needham 2007b: 117) accomplishes the land readjustment. It is a municipal readjustment committee, assigned to the municipality. After the designation of the readjustment area, the readjustment committee virtually merges all properties within the affected area into one bulk of land (Davy 2007: 41). The reassignment of the property right to individual landowners results from the value of the properties. Each landowner is entitled to receive a plot, proportionally as valuable as the original plot (in relation to the value of the bulk of land). The bulk usually increases in value due to readjustment. So, each landowner receives more than he brought in. Those who do not receive a plot for any reason, receive at least an appropriate compensation. The municipality may claim the value increase due to readjustment. Finally, the readjustment plan confirms the new allocation of building plots, and determines rights and obligations of all involved parties (Davy 2007: 41).

German law determines that in case of a land readjustment municipalities are entitled to retain some land from the bulk for public purposes (Davy 2007: 46). This makes land readjustment a usual instrument of land management to make land available for public purposes; land mobilisation for retention measures is also subject of land readjustment (e.g. see UBA 2003: 121).

However, retention measures for retaining extreme floods require extra-large areas. Administrative borders would hinder land readjustment for LATER: the required administrative level needs to be at least regional, but regional planning has no competence to intervene in the property rights in this manner. In addition, municipalities usually become the owner of the land they gather from land readjustment, which is not the LATER way. But land readjustment effectively, efficiently and fairly prepares land for development (Davy 2007: 42). However, German mandatory land readjustment is not applicable for LATER, but it is a good starting point to develop a land policy.

Mandatory protection readjustment   In case of extreme floods, flood protection is a scarce good. This scarce good is currently distributed among the whole catchment area. Both, the downstream and the upstream party receive almost the same flood protection level. LATER reallocates flood protection: a valuable downstream party receives more flood protection through retention upstream. The upstream party, evidently, has a reduced protection against inundation. Moreover, the inundation is then even wanted in case of extreme floods upstream.

As explained, mandatory land readjustment makes land suitable for the designed land use, which was not suitable before (according to size and shape of the sites), produces a benefit, and uses this benefit to provide public services. LATER aims at protecting particular suitable sites against extreme floods. It produces additional profit for these landowners. Other sites upstream, however, get less protection. The profit of the downstream party serves as compensation for upstream sites. These upstream sites provide a service for downstream landowners: they retain the next extreme flood wave. In order to increase the level of flood protection downstream flood protection measures (levees) need to be reallocated. Gains and losses of this reallocation must be redistributed. Can mandatory land readjustment reallocate flood protection?

Land readjustment relies on appraisable property rights. Since property values can be appraised, gains and losses can be calculated, payments can be initialised. Land readjustment rearranges the whole property. Reallocating flood protection in the meaning of LATER, however, does not reallocate properties, it even does not reallocate levees or structural measures. Probably, only little constructional measures are necessary to implement LATER, but LATER requires an essential reallocation of protection rights. Only a stick of the bundle of rights is needed. However, an individual right for flood protection does not exist. Landowners may sue water management agencies for not maintaining an existing levee, but landowners may not sue water managers for not building or heightening a levee, at least not for extreme floods. In 2005, the Federal Flood Control Act introduced a threshold of a centennial flood. It is not apparent yet whether courts will face cases in the future, where landowners may claim for protection up to this level. Certainly, no landowner may claim protection against extreme floods, as already mentioned earlier. But what would an additional protection right, reallocated through land readjustment as proposed, be worth, if it is not enforceable? A right, which is not recognised by a court, is worthless (and is in fact no right) (Bromley 1991: 35). Flood protection must become an enforceable right to become subject of a mandatory readjustment.

What is then the advantage of such a readjustment? Land readjustment is the only instrument that produces an additional benefit for private citizens by reallocation of rights in land and allows to collect the benefit in order to spend it on a certain purpose. Some other planning instruments are indeed restrictive, take several property rights, and compensate landowners for losses, but most of them burden the public budget for the compensation. No other instrument than land readjustment (whether urban *Baulandumlegung*, or rural *Flurbereinigung*) or to some extent Urban Redevelopment Measure (*städtebauliche Sanierungsmaßnahme*) (§ § 136–71 BauGB) redistributes planning gains as well as losses. The Urban Redevelopment Measure, however, is a very specific planning instrument, only applicable in specific cases, determined in the law. Land readjustment is the only chance to implement a reallocation of flood protection with a redistribution of gains and losses as proposed.

Implementation   What is necessary to accomplish such a readjustment of flood protection? Besides several administrative and legal adaptations, implementing mandatory protection readjustments requires three essential phases.

In the first phase, the state assigns individual flood protection rights: the key difference to traditional land readjustment is that not land, but protection rights, are subject of readjustment for LATER. Such rights do not exist:

> A right is the capacity to call upon the collective to stand behind one's claim
> to a benefit stream. Notice that a right only have an effect when there is some
> authority system that agrees to defend a right holder's interest in a particular
> outcome (Bromley 1991: 15).

Daniel W. Bromley emphasises that property is not the real land, but the income stream, which is protected by the state. Furthermore, he notes that rights are not a relationship between a person and an object, but rather a relationship between a person and others with respect to that object (Bromley 1991: 15).

If rights determine relationships between people, to whom are the flood protection rights assigned initially? Following the Coase Theorem:

> the initial assignment of property rights does not matter as long as volitional bargains are possible; the rights will go to the highest bidder and efficiency will be assured regardless (Bromley 1991: 36).

Two conditions must therefore be fulfilled: first, transaction costs – costs of information, bargaining and enforcement – are negligible; second, the wealth effects of alternative rights assignments are negligible, too. Daniel W. Bromley neglects this assumption. The theorem by Ronald Coase for Bromley seems like a tautology, and he argues that it has no relevance for the real world (Bromley 1991: 36). So, it does matter to whom the rights are assigned initially.

To readjust protection rights, they have to be privately owned. Only then, individual gains and losses can be calculated, and compensation can be claimed:

> [Private ownership is] in every society, the right to use of at least some resources ... that is, exclusively and voluntarily transferable. Under this arrangement, the owner of the right has the exclusive authority to decide how the resource is used given a set of permissible alternatives (De Alessi in Hill and Meiners 1998: 9–10).

This means that landowners may sell protection rights, or claim for compensation if others violate these rights. How can others violate these rights? Whenever one person enjoys a right, a duty of other people is associated with the rights (Needham 2006: 31). Does, for instance, an upstream person violate the protection rights of downstream owners if the upstream builds single-family houses in the floodplains (which reduce the retention capacity), or does a downstream violate the rights of an upstream by building in the floodplains (which requires additional retention upstream)? Assigning such flood protection rights is accompanied by many crucial legal questions. However, the initial assignment of rights for protection against extreme floods to individual landowners is essential for accomplishing mandatory protection readjustment. The readjustment committee can only readjust rights respected by courts. The first essential phase for implementing such mandatory protection readjustments is the initial assignment of protection rights to landowners, in order to readjust them in the second phase.

In the second phase, the readjustment committee reallocates rights. Benjamin Davy (2007) describes the formal procedure of German mandatory land readjustment in five steps. The formal commencement of the procedure identifies properties in the readjustment area, in the preparation of readjustment the individual shares of each landowner are evaluated, then value capture and reallocation takes place to calculate compensations, the readjustment plan is compiled then, before the final step – the implementation – completes the land readjustment (Davy 2007: 41).

Mandatory protection readjustment requires similar steps for allocation: initially, the readjustment area needs to be determined. This area has to contain the upstream emergency retention area as well as the downstream area. Both areas need not necessarily be spatially connected. Probably, the retention area lies far upstream to the regarded downstream area. In contrast to traditional land readjustment, the readjustment committee must then act on an regional but probably even transregional or transnational level, which indeed would increase coordination enormously. The readjustment area must consist of cohesive valuable downstream sites and considerably less valuable and also cohesive upstream sites. The committee must identify all properties within the readjustment area, which is challenging even in the well developed system of German land cadastre. Then, appraisers must evaluate the value of the protection rights to calculate individual shares of the calculatory mass. The committee reallocates flood protection rights wisely. Probability of inundation, hydraulics and potential damages is information required for this reallocation, which must achieve an increase in efficiency of the allocation of flood protection. So the committee reallocates the rights based on expertise.

In the third phase of implementation, the committee determines payments: the committee can profit from the well-elaborated proceedings of land management with respect to the distribution of gains and losses. The distribution ratio (*Verteilungsquotient* 'q') for example helps to calculate the compensations. It expresses the total value increase by dividing the value of the reallocated properties through the value of the original properties. If this ratio is bigger than one, a benefit increase has been realised. Landowners are entitled by German law to receive plots that are at least as valuable as their former plots (§ 56 BauGB). The individual share of each landowner is multiplied with the distribution ratio. The result is the portion of the bulk each landowner should – at least approximately – receive in order not to be disproportionately advantaged or disadvantaged. Readjustment for LATER virtually aims at redistributing plots which are different to the former plots: according to the level of flood protection, the downstream landowners receive plots that are more valuable; upstream, the retention areas receive less valuable plots (with a reduced protection level against flooding). The distribution ratio helps then to calculate how much money landowners have to pay for the additional flood protection, and the compensation for landowners who receive less flood protection. The committee finally induces and controls the payments.

This is a very rough description of the hierarchic land policy. It has to challenge several issues. Some came up in a discussion with members of the *Hochwassernotgemeinschaft Rhein e.V.*, an initiative of municipalities, politicians and civil action groups. It is not yet clear how politically sensitive land uses can be treated within this land policy like (e.g. a Jewish graveyard in the flood-prone area). Politically, it makes a difference whether this graveyard will be inundated by an extreme flood by fate, or whether it will be sacrificed in order to protect an industrial area downstream regardless of any monetary compensation. However, it will be not discussed more in detail yet.

Land policy implications The strong role of municipal and governmental authorities within hierarchic land policy burdens the implementation to the public. LATER becomes a public challenge. This has advantages: the readjustment committee could try to pursue other aims besides flood protection as long as these other issues do not interfere with the original task to improve the efficiency of the allocation of flood protection measures against extreme floods. Probably, land thrift or environmental goals might be issues, which could be combined with LATER. This, however, is a risk as well. Challenging the committee with too many simultaneous tasks requires a wise balancing of all issues. But the committee should not replace planning. It just has to achieve less damage in case of extreme floods. Nonetheless, allocative and distributive effects are not negligible.

The reallocation of the protection rights distracts water-sensitive uses (e.g. chemical industries, do-it-yourself stores, residential areas) from the upstream retention areas, but it sets incentives for these uses to be allocated in the protected downstream areas. Probably, water-sensitive uses are reallocated to flood-secure places. This might mean that such uses do not take place in particular regions at all. Poor people remain in the upstream areas, because they cannot afford to build on the valuable protected downstream land. Allocative aspects are fundamental consequences of flood protection readjustments.

Distributional aspects are also crucial: landowners of land upstream to big cities or industries, which was just normal rural land before, receive an additional land rent from allowing inundation of the land. Indeed, the land is less easy to use for traditional land uses: for instance, residential areas are only possible if houses are adapted to flooding. Incentives to develop such land for intensive urban development are reduced. Why should landowners bother finding investors, getting building permissions and maintaining structures, when land rent is available from just keeping the land as it is and allowing it to be flooded occasionally? As mentioned above, probably the poorer landowners indeed remain in the retention areas. But they receive money from the valuable land users downstream, who profit by the retention. These are interesting distributional aspects of the mandatory land readjustment.

In the end, the land policy leads to a risk adapted land use and a mitigation of the risk. Nonetheless, an interesting question remains according to landowners in the downstream area who cannot afford the payments the readjustment committee appraised for them. Land policy by mandatory protection readjustment is a land policy for the majority, not necessarily for the poor.

Mandatory protection readjustment is a hierarchic land policy. Hierarchic refers to the rationality, which Cultural Theory describes. The characteristics Michel Schwarz and Michael Thompson assign to the rationalities match (compare with Figure 3.1). Mandatory protection readjustment is a prospering idea in a social context of a strong structural gird and positive group adhesion. The strong structural grid is necessary for establishing such an administrative complex procedure, which will require nested bounded group organisations. The land policy strives to maintain public control over flood protection; the readjustment committee anticipates the best welfare for the most with the help of determined appraisal methods. Each landowner is equal before the law and may go to court if he feels disadvantaged.

Composed: Business as Usual LATER

Fatalists keep relaxed. As they believe in the uncontrollability of the world, they do not aim at a specific land policy for LATER. This does not mean that this rationality is indifferent. But fatalists believe that they cannot influence what happens. Pursuing a specific land policy, whether by privatisation, land trusts or readjustment, involves an expectation that the pursued approach influences the world. But owing to their belief, only one strategy remains: keeping composed.

Fatalists react to their environment in a minimum scale according to time and space. End-of-pipe technologies result from this. Levees are such a technology. In this manner, they accelerate the flood wave and increase flood risk. Fatalists are aware of this, but they believe that even LATER cannot control the risk. In 2002, the unintended inundated areas along the River Elbe were the large areas for temporary emergency retention – without being prepared for it.LATER in this situation, it is the best to use the floodplains for as long as possible, as profitably as possible. Short-term profit is the best each landowner can get. Long-term planning is impossible because nature is capricious. How does such a composed land policy emerge?

First, landowners do 'business as usual' with the floodplains. Since floodplains are inconspicuous, landowners settle in floodplains and build houses, regardless of the risk.

After flood events, fatalists react. In the short-term, they protect the values against the next flooding by levees. LATER is not desirable, because its effect is not predictable. Thus, fatalists enforce and heighten levees.

Then fatalists continue in exploiting the floodplains. 'Don't panic' is the fatalistic advice, like Douglas Adams wrote in *The Hitchhiker's Guide to the Galaxy*. Composure is the most essential characteristic of this particular floodplain management.

Neither will LATER be allocated efficiently nor will the gains and losses be distributed fairly (because fatalists do not believe in fairness). Implementing the fatalistic approach, the individuals have to be resistant against fate – in particular against catastrophes. Therefore, composure is the essential attribute. Exploiting floodplains will only be successful if the individuals do not care about uncertain risks.

Comparison

The constitutional, the competitive, the cooperative and the composed land policy for LATER differ essentially from each other. Figure 3.2 shows an overview of the four land policies: the individualist's competitive land policy, the egalitarian approach to a cooperative land policy, the constitutional land policy of the hierarchic rationality, and the composed land policy of the fatalists. Each land policy is rational from the perspective of one rationality; no land policy is rational from the perspective of all rationalities.

Figure 3.2 reminds that all proposals for land policies are derived from a prototype. Land policy is an intervention in the allocation of land and the distribution

	Individualists 	Egalitarians 	Hierarchists 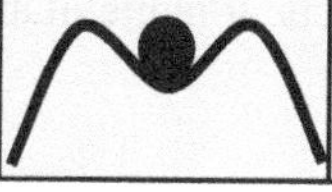	Fatalists
	CONCEPT			
Land Policy	Competitive	Cooperative	Constitutional	Composed
Prototype	The lake and the fishermen	Land trusts for land conservation	Mandatory land readjustment	End-of-pipe technologies
Idea	Concessions for flood protection	Risk alliances by land trusts	Readjustment of protection-rights	Profitable use of floodplains
	IMPLEMENTATION			
Intervention	State offers concessions for flood protection	State withdraws responsibility for flood protection	State assigns individual flood protection rights	State and municipalities do business as usual
Allocation	Concession-holders search for efficiency	Landowners establish land trusts	Readjustment committee reallocates rights	State and municipalities support levees
Distribution	Stakeholders negotiate about payments	Land trusts negotiate about payments	Committee determines payments	Landowners exploit the floodplains

Figure 3.2   **Comparison of polyrational strategies**

of the gains and losses from this allocation. The three phases of implementation show how each land policy intervenes, allocates and distributes.

Land policies differ not only in their rationality, but also in other aspects. One important aspect is the distribution of initial transaction costs. So far, transaction costs were not taken into account. The actual costs are difficult to estimate, but for the application of the land policies, it is crucial who initially has to provide money for LATER.

The strategies are not only different in the distribution of transaction costs, but also in their inherent concept of justice. A choice of a certain land policy is also a choice for a certain concept of justice. Choosing a concept of justice has two critical aspects: it is inevitable, and it is always the subject of criticism (Davy 1997: 255).

Transaction costs   Three types of transaction costs exist: enforcement costs, information costs and bargaining costs (Bromley 1991: 36). Primarily, enforcement costs are incurred for the intervention of each land policy. Finding efficient allocations for LATER requires detailed information about the areas. Identifying efficient allocations of LATER mainly produces information costs. Bargaining costs are costs of negotiating payments, respectively appraising compensations and fees. This type of transaction cost can be assigned to the distributional aspects of the four land policies. In final consequence, the types of transaction costs are related to the three phases of implementation. Figure 3.3 illustrates who pays the transaction costs.

In the individualistic land policy, the water management agency has to bear the expenses of preparing the public tender and staging the competition. The withdrawal of the public responsibility for protection against extreme floods costs nothing, but politicians end up paying some kind of transaction cost: they lose power and influence. They can compensate this loss by informing the public about the reasons and background, however this also costs money. These efforts are transaction costs for initiating this land policy. The government takes the transaction costs of assigning protection rights to the landowners for the protection readjustment. These costs represent the effort of discussing the details of the assignment in the parliament and implementing it in the administrative system and costs for enforcing flood protection rights. The fatalists in their first step produce no transaction costs, because they do not transact anything.

The decisions about the allocation of LATER produce information costs. The deciding stakeholder bears the costs. The potential concession-holders investigate in the catchment area in advance of their bid to calculate the offer. These costs are not remunerable. They are transaction costs. Finding appropriate land trusts requires a lot of information, too. Only with such information, appropriate partners for land trusts can be found. The readjustment committee must gather information on the risk in the catchment, hydraulics, values, etc. to identify appropriate allocations for LATER. The transaction costs in the fatalistic land policy remain with the inundated landowners. The damages are the transaction costs.

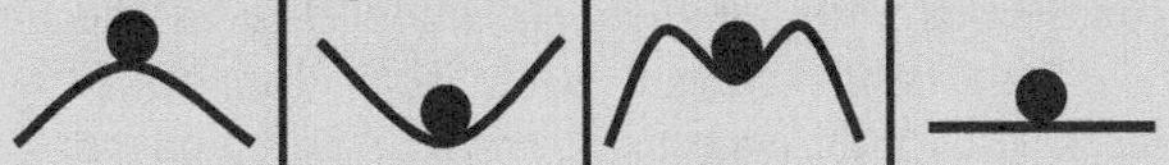

	Individualists	Egalitarians	Hierarchists	Fatalists	
enforcement costs	State offers concessions for flood protection	State withdraws responsibility for flood protection	State assigns individual flood protection rights	State and municipalities do business as usual	**intervention**
	Water-management agency	*Politicans*	*Government*	*No costs*	
information costs	Concession-holders search for efficiency	Landowners establish land trusts	Readjustment committee reallocates rights	State and municipalities support levees	**allocation**
	Potential Concession-holder	*Landowners*	*Readjustment committee*	*Inundated victim*	
bargaining costs	Stakeholders negotiate about payments	Land trusts negotiate about payments	Committee determines payments	Landowners exploit the floodplains	**distribution**
	Concession-holder	*land trusts*	*Readjustment committee*	*No costs*	

Transaction-costs of implementation

Figure 3.3   Transaction costs of the land policies

The distribution of gains and losses are bargaining transaction costs. This enforcement costs money, which are transaction costs. Costs for negotiating with landowners burden the concession-holder in the hierarchic land policy respectively the land trusts in the egalitarian land policy. Readjustment committees have to enforce the individual payments of the landowners. In the fatalistic land policy, there emerge no relevant transaction costs for exploiting floodplains.

Finally, the initiating stakeholders in each land policy pay transaction costs for enforcing the land policy. Costs for information about the efficient allocation of LATER are most risky for the stakeholders, because in each land policy these costs might not pay off, if an appropriate allocation cannot be found, or – in the individualistic land policy – the concession-holder loses the competition. The bargaining transaction costs for the distribution of gains and losses are to the expense of the winners of each land policy. This allows the remuneration of the transaction costs through the net benefit, which LATER produces. Barrie Needham emphasises that:

> if the aim of that planning (or at least one of its aims) is to use economic resources efficiently, then according to the Coase Theorem (as it has come to be called) that will be achieved by actions which minimize transaction costs (2006: 60).

Finally, the implementation of the land policies depends on identifying efficient allocations cheaply.

Concept of justice   LATER achieves Kaldor-Hicks efficiency. From an economic point of view, this efficiency is still desirable if winners are already rich and losers already poor (Needham 2006: 64):

> Laws and institutions no matter how efficient and well-arranged must be reformed or abolished if they are unjust (Rawls 2005: 3).

Efficiency does not convince stakeholders if they perceive injustice (Davy 1997: 282). For the land policies, justice is crucial.

Different concepts of justice exist: libertarian (or elitist) justice, social justice and utilitarian justice (Davy 1997: 257). Individualists prefer minimal state in order to let the strong prevail; the invisible hand of the market leads to libertarian justice, which emphasises the liberty of individuals. Social justice advocates the poor, and promotes the welfare state; it is a concept of justice, which is assigned to the egalitarians. Utilitarian justice pursues the key axiom 'maximise happiness', which benefits the majority; state is the keeper of justice, it is the favourite concept of justice of the hierarchic rationality (Davy 1997: 267, Davy 2004: 166). Fatalists do not believe in justice, consequently, they have no concept of justice (Schwarz and Thompson 1990: 66–7). Figure 3.4 shows the four land policies and their approach to justice.

Each land policy creates different winners. The competitive land policy pursues libertarian justice. The concession-holders are the winners in this idea of a land policy. Indeed, each landowner benefits from the land policy, but the concession-holders are the winners who gather money from LATER. Only the strongest, the most efficient concession-holder receives this advantage. He is then also the executor of justice (with some constraints through the public agency who tenders the concession and the legal framing in which the privatisation of flood protection of extreme floods takes place).

Single landowners profit from the land policy by LATER land trusts. Land trusts like the here proposed ones also enable poorer landowners to become powerful stakeholders, because in these trusts landowners are reliant on each other. This

	Individualists	Egalitarians	Hierarchists	Fatalists
Prototype	The lake and the fishermen	Land trusts for land conservation	Mandatory land readjustment	End-of-pipe technologies
Idea	Concessions for flood protection	Risk alliances by land trusts	Readjustment of protection-rights	Profitable use of floodplains
Winners	Concession-holder	Landowners	The majority in the catchment	Randomly
Decision-makers	Concession-holder	Land trusts	Readjustment committee	Nobody
Concept of justice	**Libertarism**	**Social justice**	**Utilitarism**	-

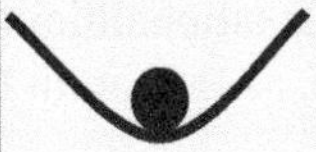

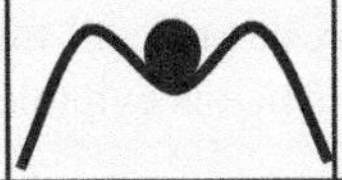

Figure 3.4   Concepts of justice

benefits poor landowners, because they gain power without investing money. The land trusts distribute gains and losses.

The constitutional readjustment of protection rights aims at increasing the welfare for the most, because the readjustment committee is a public institution, which must be democratically legitimated by an increase of benefit for the majority. The readjustment committee has the task to make the benefit available for this majority. Thus, it is the executor of justice in this utilitarian land policy.

Owing to a lack of belief in justice, fatalists pursue no specific concept of justice. Who profits most is just a random occasion of the next extreme flood. Some become victims, others profit by crevasses upstream.

Choosing one land policy is a decision for one concept of justice against the other concepts. As long as one land policy would be applied in a world where all inhabitants of this world adhere to the same concept of justice, the choice would be an easy one. But if at least two different concepts of justice are desired by a relevant number of people in this world, the choice means to respond to one group by neglecting the other one. As discussed earlier, the land policies manage the clumsy floodplains. Clumsy floodplains however are far from adhering one concept of justice, because all four rationalities are involved in the clumsiness. In this situation, injustice is evitable, as Benjamin Davy emphasises (1997: 282–3).

Contesting Land Policies

In clumsy floodplains, land policies contest with each other. Clumsy floodplains are a result of a social construction. Four stakeholders are essential for this social construction: landowners of the floodplains, water management agencies, policymakers and land use planners. In four phases, these stakeholders perceive floodplains as profitable, dangerous, controllable and inconspicuous. These perceptions result from rationalities, and adhere to the situations in the phases. The legal system sustains the rationalities. The rationalities permanently contest about determining the actions in floodplains. In each phase, one particular rationality dominates the situation in floodplains. Finally, clumsy floodplains are in a permanent dynamic imbalance of rationalities.

The proposed land policies aim at LATER. LATER is floodplain management, which is able to cope with extreme floods. Extreme floods virtually swamp the capacity of flood protection. By reallocation of flood protection, the efficiency of flood protection measures against extreme floods can be increased. This reallocation requires redistribution, because flood protection is a scarce good almost at its production possibilities frontier. Despite LATER being more efficient than contemporary flood protection, the market is not able to achieve it. The social construction of floodplains supports the levee-based flood protection approach. In the future, floodplains will become technologically locked-in to levee-based flood protection. The land policies must initiate an optimal perturbation in the clumsy floodplains in order to overcome the technological lock-in.

In a monorational world, the four land policies – except the fatalistic approach – are theoretically able to achieve LATER. But the clumsy floodplains are not monorational. The monorational land policies have to challenge with other rationalities. These other rationalities undermine monorational land policies.

Polyrational Land Policies in a Polyrational World

Social and political life is:

> an ongoing competition between groups of people who adhere to one of these different cultures (Verweij 2000: 5).

How does each of the monorational land policies stand in the polyrational arena? This question depends on the point of view of the other rationalities to the rationality regarded. Each of the rationalities recognises the others and has an opinion about

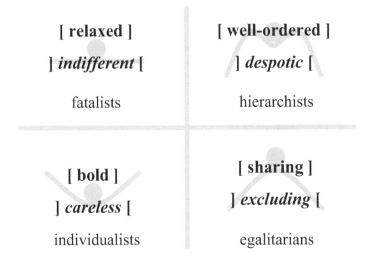

Figure 3.5 **|Internal| and |external| perceptions**

the others. These opinions differ from the perceptions each rationality has about itself (whereas it is difficult to assume that situations 'perceive', but it is suitable for the moment to personalise rationalities). So, what are the internal and external perceptions of the rationalities, and what effect do these perceptions have for the contest of the land policies in the polyrational world? The following sections deal with these questions.

Internal and external perceptions   The internal and the external perceptions of the rationalities differ essentially. Individualists perceive themselves as bold and brave, while others call them careless. Egalitarians are cooperative and sharing from internal, but excluding from external, perception. The hierarchic rationality is perceived from an internal point of view as well ordered, whereas other rationalities perceive hierarchists as despotic. The fatalists perceive themselves as relaxed; the others perceive them as indifferent. These internal and external perceptions are summarised in Figure 3.5. The own perception has a positive connotation; the external perception has a negative connotation (Davy 2004: 279, Davy 2008: 308).

The land policies are monorational solutions for a polyrational situation. Deriving from the internal and external perceptions, each land policy is negative from three rationalities. Figure 3.6 shows the perception of each rationality for each land policy in columns; the missing properties from the perspective of each rationality are in the rows. For instance, hierarchists evaluate the readjustment land policy as well ordered. Hierarchists are satisfied with this land policy. Regulation is missing in all the other land policies: the privatisation is careless land policy, land trusts are exclusionary from the hierarchists' perspective, and the fatalistic

		P e r c e p t i o n of the land policy		
		Concessions for flood protection	**Risk alliances by land trusts**	**Readjustment of protection**	**Profitable use of floodplains**

	Individualists	**bold**	*excluding* *bold*	*despotic* *bold*	*indifferent* *bold*
	Egalitarians	*careless* *sharing*	**sharing**	*despotic* *sharing*	*indifferent* *sharing*
	Hierarchists	*careless* *well-ordered*	*excluding* *well-ordered*	**well-ordered**	*indifferent* *well-ordered*
	Fatalists	*careless* *relaxed*	*excluding* *relaxed*	*despotic* *relaxed*	**relaxed**

*(Row axis label: **D e m a n d e d** land policy-properties)*

Figure 3.6   Perception matrix

approach is too indifferent. Each rationality can reason why land policies from another rationality are not appropriate, and each rationality can express what attitude is missing in the other land policies. Only land policies in accordance with the own rationality are suitable.

Killing LATER Since each rationality perceives other rationalities' land policies as negative, the described external perceptions of the rationalities lead to modifications of monorational land policies in polyrational situations. Rationalities undermine monorational land policies.

Individualists, for instance, start to introduce market schemes in a hierarchic land policy. As they perceive the hierarchic land policy not only negative, but also in fact unjust, they will start to manipulate the readjustment land policy by introducing a market scheme, probably by corruption. The consequence is that distrust in the

		L a n d p o l i c y		
		Concessions for flood protection	**Risk alliances by land trusts**	**Readjustment of protection**	**Profitable use of floodplains**
R a t i o n a l i t i e s	*Individualists*	*No modification*	*Misuse of trust*	*Agencies will become corrupt*	*Rapid exploitation*
			Land trusts fail	Distrust emerges	Risk becomes worse
	Egalitarians	*No cooperation with companies*	*No modification*	*Collective resistance*	*Collective protest*
		Negotiations fail		Paralyse agencies	Obstruction of profitable use
	Hierarchists	*Regulate competition*	*Control land trusts*	*No modification*	*Enforce plans & order*
		Obstruction of competition	Distrust in associations		Plans will not be accepted
	Fatalists	*Neglect negotiations*	*Neglect participation*	*Neglect authority*	*No modification*
		Impossibility of negotiations	Land trusts cannot establish	High costs for enforcement	

Figure 3.7   Land policy modifications

land policy emerges. The hierarchic land policy is not able to work effectively and achieve efficiency. For each combination of strategies and rationalities, certain behaviour, with certain effects on the ability to make LATER available can be estimated. In principle, each rationality destabilises key mechanisms of the other strategies. Figure 3.7 shows possible scenarios on how the rationalities might react to the land policies, and what the probable effects for the land policies are. These manipulations will finally make an implementation of LATER impossible. Polyrationalities are killing monorational LATER land policies.

Responding Polyrational

Can the other rationalities be drowned in order to implement monorational land policies? The implementation of a land policy in a world with fewer rationalities would be easier. Assume, for instance, only *homo economici* would exist in floodplains, then the individualistic land policy could be implemented without the contest against the other rationalities. If only hierarchists were supported by the legal system, the other rationalities might disappear and clumsiness would be abolished. Why does clumsiness emerge? How can policy respond to this clumsiness so that it does not 'kill LATER'?

Essential clumsiness   Cultural Theory states that situations are inconstant if one rationality dominates, for example in the police state, in the pure market economy, in the non-violent commune, in the isolated state (Davy 2004: 145, Davy 2006: 81). This is the impossibility theorem: none of the four rationalities can become permanently uninhabited (Thompson et al. 1990: 86). If one rationality is temporarily missing, it will emerge and change the situation.

This can be observed in almost every social situation. Imagine for instance a public swimming pool, full of swimmers who want to do sports. Each swimmer wants to complete lanes without bumping into each other. Each swimmer has a certain rationality, which influences his strategy to avoid collisions. Hierarchists expect some system, some rules; like for instance that swimmers must swim on the right side of the pool in one direction and swim back on the left side of the pool. If such rules do not exist, hierarchists establish them (the hierarchists are the organisers who advise others to behave in this or that way, and who are complaining if other swimmers do not obey these rules). Individualists, on the contrary, believe in a contest about lanes. Swimming is a competition; the lanes in the pool belong to the best swimmers (they are the reckless swimmers who drown the weak). Furthermore, there are the egalitarian swimmers. For them, swimming is a social interaction with others; they go swimming to meet others. They swim carefully and give way to other swimmers – but they do swim almost in a zig-zag course. Fatalists, however, just jump into the water and swim, convinced that bumps are not avoidable. They swim their lanes, and probably they bump into others or not. This swimming pool is full of different rationalities. In sum, the movements look a bit awkward. Indeed, most swimmers complete their lanes more or less, but several swimmers bump into each other. If all swimmers would follow the same rationality, the swimming in the pool would seem less clumsy. Nonetheless, in most crowded public swimming pools clumsy swimming can be observed.

Each bump causes a surprise for the swimmers. Each rationality expects the other swimmers to behave correspondingly. A bump surprises the swimmer, because it does not fit into the expectation of the world (for those who expected bumps, like fatalists, the absence of bumps surprises) (Thompson et al. 1990: 70). The swimmer reconsiders his perception and expectation of the world – of the swimming pool. He assumes then that swimming in the pool follows another rationality. He changes rationality. In the end, the rationalities in the swimming pool are in a 'permanent dynamic imbalance', where no rationality remains uninhabited, but individual swimmers permanently change their rationality (Thompson et al. 1990: 69–72, 87).

Assume that for some reason all adherents of one rationality leave the pool – e.g. a stock exchange crash pulls all individualists to their bank service desks, or all hierarchists obey some rule that swimming is not allowed after 7:30 p.m. – however, all adherents of one rationality leave, but the pool is still quite full. Due to time, the remaining swimmers will fill the gap and change rationalities until each rationality is again present in the pool. The reason for this is the typology of surprise (Thompson et al. 1990: 69–72).

Excluding one rationality permanently from the clumsy floodplains is impossible. Rationalities cannot be drowned permanently. Clumsiness emerges. For that reason, monorational land policies, which are rational only for one rationality, have to cope with polyrationality. Such monorational land policies are ideal from the internal perspective of the rationality regarded. Idealism, however, undermines itself: individualism needs hierarchism to enforce contracts and repel enemies; egalitarianism needs also hierarchism, to get work done and settle disputes; hierarchism, vice versa, needs individualism to make progress, egalitarianism for cohesion, and fatalism to remain stable (Thompson 2008b: 39, Verweij et al. 2006a: 7). No rationality has exclusive access to the truth about the composition of the world. Rather the world is socially constructed by all four rationalities (Schwarz and Thompson 1990: 13):

> No solidarity, we should note, has it all its own way … but each gets a lot more than nothing. This clumsy institutions do more than just nurture trust and consent; it is also a learning system that is set up in such a way that it can learn from each of the three ways of knowing that are built into it (Ellis and Thompson 1997: 212).

The best tools to understand this inchoate are diversity, contradiction, contention and criticism. Rationalities need each other to stand:

> Divided we stand, united we fall (Schwarz and Thompson 1990).

This speaks even for clumsiness as a desirable concept. Clumsiness encompasses, as much as possible, wisdom and experience from other rationalities (Thompson 2008b: 25), despite that such wisdom and experience from other rationalities might be uncomfortable. So, Cultural Theorists claim 'May our futures be clumsy' (Verweij et al. 2006b: 247). But clumsiness by design – 'how on Earth do you do that?' (Thompson 2008b: 12).

Accessibility and responsiveness Deadlock and congestion describes the technological lock-in; policy is entrenched in the contemporary levee-based approach to flood protection. Too little democracy and pluralism lead to such deadlock and congestion, to the technological lock-in. Clumsy policy solutions increase democracy and pluralism because it listens and responds to all rationalities (Ney 2007: 316). So, designing clumsy policy necessitates access to all rationalities and response to all rationalities. It means to design a polyrational policy.

By granting all rationalities access to the policy process, pluralism emerges. The more accessible a policy is for the rationalities, the clumsier the policy will be. The degree of accessibility expresses to how many rationalities a policy grants access. Following Cultural Theory, only the three active rationalities: hierarchists, individualists and egalitarians, in fact use such access. The fatalists, as passive rationality, do not even want to raise their voice in political processes. So, a triangular policy – where all three active rationalities have access – is inclusive; a bi-polar policy is partly inclusive, whereas a policy that denies access to the other rationalities is exclusive (Ney 2007: 322). Steve Ney emphasises:

> The more social solidarities are present in a policy subsystem, the larger is the pool of the organizational, cognitive, and practical resources available for dealing with complex and uncertain policy issues ... Moreover, as accessibility increases, the risk of conceptual blindness, surprises, and policy failure decreases (2007: 322).

Accessibility increases the deliberative quality of a policy. Responsiveness is a measure for popular control. Responsiveness increases the democratic quality of a process. It describes how interactions in policy processes are structured and regulated:

> As deliberation becomes more responsive, each voice in the policy debate becomes more audible, clearer, and more sensible to contending policy actors (Ney 2007: 324).

In other words, responsiveness expresses the number of rationalities to which policy responds. Three degrees of responsiveness emerge in accordance with Cultural Theory: assertive deliberation is the lowest degree of responsiveness. Policymakers 'aim to assert their particular policy story over rival stories' (Ney 2007: 323). Strategic deliberation is the next degree, where two rationalities respond to each other to achieve a more effective policy. Reflexive deliberation, finally, is the highest degree of responsiveness. Actors scrutinise critically on means and ends of policymaking (Ney 2007).

If the policy responds only to a few rationalities or if only a few have access to the policymaking process, the deliberative quality is lower. In return, the more rationalities are involved in a policy – no matter which rationalities these are – the higher is the deliberative quality of the policy (Ney 2007, Thompson 2008a: 13). If all rationalities have access, and are responded to, the policy is clumsy; if only one rationality is heard and responded to, the policy is a 'closed hegemony'. Closed hegemony means that it is a monorational, and from the perspective of the regarded rationality ideal solution. Between these two extremes, several degrees of responsiveness and accessibility are possible. Steve Ney identified a scheme of nine distinctive political qualities (see Figure 3.8); the scheme shows the 'regions of pluralist democracy' (2007: 236). He characterises each of the

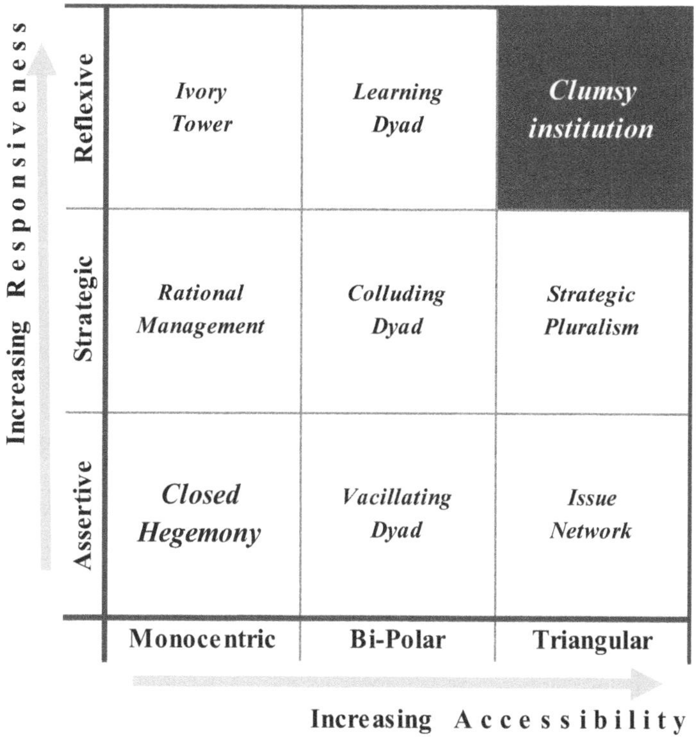

	Reflexive	*Ivory Tower*	*Learning Dyad*	*Clumsy institution*
Increasing Responsiveness ↑	Strategic	*Rational Management*	*Colluding Dyad*	*Strategic Pluralism*
	Assertive	*Closed Hegemony*	*Vacillating Dyad*	*Issue Network*
		Monocentric	**Bi-Polar**	**Triangular**

Increasing A c c e s s i b i l i t y

Figure 3.8   Regions of pluralist democracy

nine policy approaches (Ney 2007: 353–61). The scheme shows nine policy qualities, because the quality depends on the number of rationalities who have access and the number of rationalities to which the policy is responsive. Since only the three active rationalities are relevant for policy processes, a field of three to three emerges.

The landscape of clumsy solutions   The 'regions of pluralist democracy' shows nine policy qualities which differ not only in accessibility and in responsiveness, but which produce a different 'degree of clumsiness'. Michael Thompson put a third dimension to the scheme to show the relation between accessibility, responsiveness and clumsiness (2008a: 13). A landscape of clumsy solutions emerges (see Figure 3.9).

It can be read like a way to achieve clumsy policies: increasing either accessibility or responsiveness of political processes increases clumsiness. Each policy approach can be assigned to one of the nine fields by analysing its accessibility and responsiveness. Each policy achieves some degree of clumsiness, which is a precondition for robust policy. Michael Thompson advises:

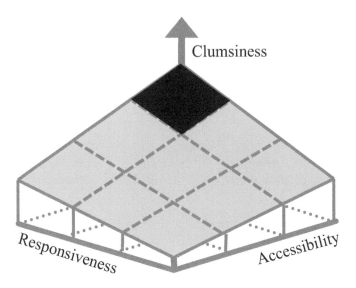

Figure 3.9   Landscape of clumsy solutions

> If you're in the midlands head for the highlands, and if you are in the lowlands
> begin by heading for the midlands. And, wherever you are, don't be cajoled into
> moving downhill! (Interview Thompson 2009).

There are many ways uphill. A policy in a closed hegemony can be 'clumsified' either
by granting access to other rationalities, or by responding to other rationalities. The
best, however, is to grant access to all three active rationalities and simultaneously
to respond to the three rationalities. The landscape of clumsiness do not regard
fatalists (or hermits):

> For policy design, however, you will need to take some account of the fatalist's
> policy input: Why bother! (Thompson et al. 1999: 13).

Excluding fatalists (and hermits) for policy discourse analyses is a simplification
for 'user friendly Cultural Theory' as Michael Thompson et al. put it (1999: 13).
Such simplifications are possible and useful. The authors of the book *Cultural
Theory as Political Science* even advise scientists to ask themselves:

> In relation to the application you are contemplating, how much of Cultural
> Theory you are going to need (Thompson et al. 1999: 13).

This does not mean that Cultural Theory is arbitrary; rather this statement points
out that Cultural Theory can be handled in a more or less flexible way in accordance

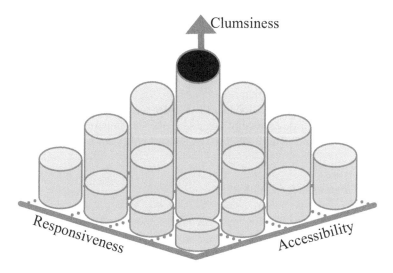

Figure 3.10   Enhanced landscape of clumsiness

to the purpose of its application. In other words, the power of Cultural Theory is not its precision, but rather its flexibility.

In clumsy floodplains Cultural Theory helps to design a viable policy for LATER, using the landscape of clumsiness requires an adaptation. Not a 3x3 matrix but rather a 4x4 matrix is needed, whereas the art of creating a clumsy policy is hearing a rationality, which does not want to raise its voice, and responding to a policy, which does not believe in policy. It has to be emphasised that the 4x4 matrix includes the possibility of a closed hegemony of a fatalistic policy (this would be a policy that does not govern at all). In addition, several fields in the matrix are not labelled yet, but that is not a key problem right now.

A problem indeed is the gradient in the slope. Michael Thompson emphasises that there are just three altitudinal zones (referring to the 3x3 matrix) (interview Thompson 2009). The gradient of the slope, however, suggests that the responsiveness and accessibility can be increased continuously and linearly. But a policy cannot grant access to, for instance, 1.5 or 2.7 rationalities. Only one, two, three or four rationalities are possible. Also, the graph pretends to have an origin, but a policy cannot include none of the rationalities (this would indeed be an odd policy). In fact, the slope depicted in Figure 3.10 consists not of a consistent plane, but rather of nine spots. So, columns represent the different stages better than a slope.

Finally, policies that respond only to one or a few rationalities and deny access to the others or a few others will be undermined by the missing rationalities until a clumsy situation emerges, no matter whether the intended policy aim will be achieved or not. Clumsiness, Cultural Theory points out, emerges in each social situation in the course of time. The surprise of the other rationalities will manipulate a monorational policy. Setting up a robust policy to achieve a long-term policy aim requires access for all and response to all rationalities. Moving a policy uphill the landscape of clumsy solutions makes it more polyrational but clumsier as well. In return, it becomes a robust policy. Disregarding the landscape of clumsiness will kill *LATER*, as described earlier.

Policy design for clumsy floodplains in order to achieve LATER requires the inclusion of the fatalistic rationality, because it plays an important role in the social construction. How can an accessible and responsive policy for LATER be designed?

A Clumsy Response for LATER

A clumsy response for LATER must encompass hierarchy, market and solidarity in one policy, according to the three active rationalities. 'Encompass' means that it must grant access and respond to the three steering approaches. Accessibility requires that some public institution must be able to control the land policy for LATER, that the market influences the allocation and distribution, and that solidarities play an important role for the implementation. The land policy must reward voluntary accomplishment of LATER to respond to egalitarians, set incentives to respond to the individualistic rationality, and determine sanctions and rules for hierarchists' obedience – simultaneously. Insurances could solve this task polyrationally.

Insurances as Land Policy

Usually, insurances are seen as institutions for the aftermath of a disaster. But insurances do have some effect on the allocation. Two opposite allocative effects are possible: on the one hand, the coverage of a potential damage reduces individual risk of losses, which pulls urban development in flood-prone areas. On the other hand, if the premiums are high enough in flood-prone areas and low enough in other areas, it can be an incentive to settle in relatively flood-secure areas (Munich Re Group 2003a: 24–5). Currently, these effects are side-effects of insurances. Probably these effects can be used more purposefully in a land policy through insurances.

Insurances are market institutions. Insurances are based on the idea of solidarity. Some insurance are legally prescribed, automobile liability insurance in Germany or the Netherlands for instance. Insurances contain constitutional, competitive, cooperative and even composed elements. In addition, insurances are already involved in coping with floods. Detailed information on flood risk costs water management agencies a lot of effort, while the insurance industry has relevant knowledge already available. For instance, the information system for flooding, backwater and intense rain contains such data. German insurance companies use this data for the calculation of insurance premiums (Munich Re Group 2003a: 24). Insurances do not only have knowledge on inundation probabilities, but also detailed information on the potential damage. Can we use this information when developing insurances as a land policy for LATER? In the USA, the 'National Flood Insurance Program (NFIP) is focused primarily on flood insurance as a means for partial recovery of losses for property owners' (Loucks et al. 2008: 541). The NFIP provides subsidised flood insurances for communities, which adopt and enforce flood mitigation, and restrict land use of the floodplains (Browne and Hoyt 2000: 292). However, despite all success, in the United States a significant number of flood losses remain uninsured. Landowners of land at lower risk are not willing to purchase coverage, which reduces the risk community and heightens the premiums for landowners at high risk, which in return neglect paying sufficient premiums (Browne and Hoyt 2000: 293 and 303). Subsidised insurances, like NFIP, also have the effect that the real risk is not reflected in the premiums, which sets fewer incentives for adaptive measures. Landowners believe in levees and do not expect extreme floods (Browne and Hoyt 2000: 297). The crucial question is whether insurances can also steer allocations and influence distributions.

Insurances Against Flooding

> [Insurances] distribute the expected long-term loss burden geographically and over time, so that the individual bears only a small financial burden per unit of time, and sufficient reserves can be built up to cover even tremendous damage caused by extremely rare catastrophes (Swiss Re 2002: 3).

Swiss Re is one of the world's biggest reinsurance companies. Currently, in the most countries the state provides – implicitly or explicitly – coverage for natural hazards (DKKV 2003: 22). Insurances can disburden the state from the increased burden of natural hazards and they are a viable tool to achieve more efficiency in flood risk management (Huber 2004: 179), which is required for LATER.

Land policy through insurances must work in the whole catchment area. Opinions whether floods are catchment-widely insurable differ. Peter Heiland points out in his dissertation on flood risk management (2002: 209) that if everybody insures himself or herself, it could be a suitable internalisation instrument.

Insurability depends on six key principles: mutuality, need, assessability, randomness, economic viability and similarity of threat (Swiss Re 1998: 7). Mutuality means that a large number of risk-affected people must form a risk community. This principle is not yet met in the flood issue, because often, only landowners who are very frequently affected by a flood insure their property against flooding. For them, the premium is very high, because the risk community is too small. An obligatory insurance increases the risk community, reduces therewith the premiums and creates mutuality. The second principle, the need, expresses that the flood must place the insured landowners in a condition of financial need. This principle is met, flooding causes enormous damage, which might even affect the financial existence of landowners. Uncertainty remains, but in principle, flood risk is relatively precisely to assess. Randomness, the fourth principle for insurability, applies for floods:

> The time at which the insured event occurs must not be predictable, and the occurrence itself must be independent of the will of the insured (Swiss Re 1998: 7).

Despite all human influence on floods, flood events happen randomly. Economic viability refers to the ability of the community to cover the future loss-related financial needs on a planned basis. A small risk community will not cope with extreme floods, so the risk community has to be large enough. Similarity of threat is essential to spread the insurance over a large community. In spite of different kinds of flood events, the similarity of threats can be assumed as given in the case of flood events. In final consequence, the analysis by the Swiss Re concludes that some key principles for insurance are not met. Nevertheless, they are convinced that the hurdles are not insurmountable (Swiss Re 1998: 7–8):

> No matter how the people at risk are persuaded to join the solidarity community, not every policyholder has to pay the same premium rates (Swiss Re 2002: 5).

The insurance companies calculate for each landowner contributions depending on the individual risk. Until now, many countries have neither systematic documentation available nor any official coordinated hazard analysis project. Thus, the insurance industry recently made investigations on their own (Swiss Re 2002: 5). On this basis, risk adapted premiums can be calculated. The premiums result on the one hand from the probability and severity of the flooding and on the other hand from the potential damage. The most risky land uses (regarding to potential damage and probability of damaging events) are the most expensive to insure:

> Premium levels, deductibles, and policyholder obligations act as incentives to avoid risk (Swiss Re 2002: 5).

Floods are insurable if risk communities are large enough. Insurance premiums can be calculated in relation to the individual risk of the insured landowners. An obligatory insurance against natural hazards can become a viable land policy.

Obligatory insurance against natural hazards The land policy for LATER is based on insurances against natural hazards. Other natural disasters, storms, wildfires, earthquakes, etc. are typically summarised in insurances against damages through inundations, earthquakes, land subsidence, avalanches and volcanic eruptions (*Elementarschadensversicherungen*) (Schönberger 2005: 193). This essentially enlarges the risk community and makes such insurance more attractive for a landowner who lives far away from a river on the top of hill. Linking several natural risks makes it possible for insurers to counterbalance frequent risks, like flooding, or earthquakes with rare risks (Schönberger 2005: 193).

Should the insurance be obligatory or voluntary? In Germany, such insurances are not obligatory yet (Schwarze and Wagner 2004: 159). The Swiss

Re points out that it 'does not necessarily entail passing new laws' to achieve an area-wide coverage with insurances (Swiss Re 2002: 3). But they promote that the government has to create the basic conditions that are indispensable for comprehensive insurance covers like promoting people's awareness of risk, mapping high exposed regions, adapting zoning plans. Furthermore, the insurers state that:

> it will hardly be possible to build up large risk collectives if the government continues to offer financial aid to people who are not insured (Swiss Re 2002: 3).

Nonetheless, in 2003 in Germany only 10 per cent of all households had contracted an insurance against flooding for personal belongings, only 4 per cent had insured their buildings as well. For historical reasons, the insurance rate is higher in Eastern Germany and Baden-Württemberg: in the German Democratic Republic, the state provided insurances against natural hazards in the whole country, such insurances were obligatory in Baden-Württemberg until 1994. The insurance sector has been deregulated since then (Schönberger 2005: 193):

> The majority of the people at risk underestimate the flood hazard (Swiss Re 2002: 4).

People have difficulties in estimating and evaluating less likely but severe events. In addition, insurers neglect insurances in some flood-prone areas like along the River Mosel or some parts of Cologne. The lack of demand for insurance against natural hazards is the key hurdle for an area-wide coverage in Germany (Schwarze and Wagner 2004: 160).

An obligation to insure could essentially increase the number of insured. In the aftermath of the 2002 flood events, an obligatory insurance against natural hazards has been discussed in Germany. The aim was to relieve the state and the voluntary aids after a flood (Schönberger 2005: 197). But the promoters could not achieve a political majority for this instrument. The enormous efforts of the state and voluntary aids after the flood abated the interest in obligatory insurances (Lange 2005: 156–7, Browne and Hoyt 2000: 297). An obligation, however, would be necessary for the land policy through insurances. The land policy works like the following.

First, the state establishes an obligatory insurance: an obligatory insurance against natural hazards is similar to the 'health insurance' in Germany. It is an obligatory insurance, which covers an individual risk. Reimund Schwarze and Gerd G. Wagner propose to apply such an insurance only to centennial floods, for the reason that very flood-prone areas, such as along the River Mosel, can be insured as well (2004: 155). Every landowner must contract insurance against natural hazards, including floods. The state must not subsidise the premiums in order not to create market failures. But the state must enforce the obligation. Insurance companies calculate individual risk-adapted premiums.

Private insurance industry has the know-how and experts necessary for reliable
assessing catastrophe risks (Swiss Re 2002: 3).

Of course, precautionary measures of landowners reduce the premiums as well as
municipal measures, whereas water-sensitive land uses and constructions in flood-
prone areas are more expensive to insure. In floodplains, premiums are very high,
in flood secure areas, the premiums – at least regarding floods – are rather low.

Second, the insurances set incentives for LATER: supported by the
municipalities, insurance companies offer landowners to reduce their high
insurance premiums if they prevent extreme floods. A study on the willingness
of Dutch landowners to invest in damage reducing measures came to the result
that insurances can stimulate such investments by offering certain benefits on
insurance policies (Botzen et al. 2009: 26). LATER is a measure that crucially
reduces the risk of extreme floods. The next extreme flood will inundate some
other area upstream. The premiums are objectively calculated and risk-oriented.
In consequence, insurers reduce premiums.

Nonetheless, mobilising LATER requires more than one landowner. Like in
the cooperative strategy, landowners establish land trusts. Then, negotiations
about LATER take place like in the egalitarian land policy. The difference is
that insurance companies provide data about risk in the catchment area to their
customers. The land trusts receive information about potential LATER areas. This
reduces the transaction costs for identifying efficient allocations for LATER in
comparison to the earlier described land policies.

In the third phase, land trusts negotiate about LATER: a downstream land trust
offers an upstream land trust payments for providing inundation rights in return. If
both parties agree, the upstream land serves as retention volume for the downstream.
The risk of the upstream landowners increases through this deal. So, the insurance
premiums become quite expensive. The payment of the downstream landowners,
negotiated through the land trusts, must cover the increased insurance premiums;
respectively, the reduction of the premiums for the downstream must cover the
payment to the upstream landowners. These two conditions are only fulfilled if a
risk alliance can be found for which the Kaldor-Hicks criterion is met. Probably,
the upstream can still reduce the premium through investing the payments from
downstream in object protection measures. The Swiss Re supports this idea:

> The insurance companies should grant premium discounts to reward structural
> measures to avoid or mitigate damage (2002: 7).

In the end, a win-win situation emerges for landowners, the state and insurance
companies: landowners are protected against extreme floods respectively profit
from selling inundation rights; insurance companies sell insurance, and the
state does not need to step in when an extreme flood occurs (landowners have
a legitimate claim on the basis of their insurance). So, an obligatory insurance
against natural hazards becomes a viable land policy for LATER.

Clumsiness of the Obligatory Insurance

Obligatory insurances against natural hazards are able to achieve LATER. By internalising risk to landowners, the most efficient allocation will be found. Redistributing gains and losses leads to the mobilisation of retention volume for extreme floods. But the monorational land policies achieve – at least theoretically – LATER as well. Is the obligatory insurance clumsy enough to intervene in the social construction of floodplains? The monorational land policies fail. Monorational land policies will not survive in a polyrational world. A viable land policy must grant access to all rationalities and respond to them.

Individualists' access and response: the individualistic rationality is heard in this land policy. Insurances are market-oriented institutions that strive for profit and efficiency. The negotiations about premiums emphasise the market and allow scope for individualists. This land policy listens and responds to the individualistic rationality. The best conditions for negotiations are after the inconspicuous phase of the floodplains, when enough time for appraisals and negotiations remains. The individualistic rationality dominates in this situation. Indeed, individualists neglect the obligation for insurance. The obligation threatens the market since it produces market failures. But as long as hierarchy is reduced to this initial step of enforcing the insurance, it will be a viable compromise for individualists as well.

Egalitarians' access and response: insurances are based on distributing the risk among the members of a large risk community. This emphasises inherent solidarity of the insurance solution, which is a response to the egalitarians. In addition, egalitarians are able to raise their voice in the establishment of land trusts for LATER. Indeed, land trusts are very powerful in the scheme. Their role satisfies the egalitarians. The best phase for establishing land trusts within the social construction is the phase when floodplains are perceived as dangerous. So, the social construction supports the land policy by insurances. Also for egalitarians, the described land policy is only a compromise, far from ideal. The strong market stakeholder, namely the insurance companies, can only be accepted if the land trusts become strong solidarities. Egalitarians also distrust the hierarchists, who set up the obligations. But the land policy by obligatory insurance against natural hazards is as much egalitarian as possible.

Hierarchists' access and response: the insurance is obligatory. This requires a legal founding and an institution, which controls the obligation. In addition, penalties are required if landowners refuse to contract an insurance. Probably, some adaptations of the obligation for insurances are necessary due to time. This requires hierarchy. The hierarchic rationality is heard and responded to in this land policy. In the social construction of floodplains, the hierarchic rationality emerges some time after a flood. This, however, is the perfect time for introducing an obligatory insurance against natural hazards, because the flood is close enough, but reconstructions etc. are already done. Nonetheless, hierarchists will not estimate the land policy by insurances being optimal. Allocation and distribution remain to market actors and civil initiatives. A perfect hierarchic solution would assign this task to an agency. However, the obligatory insurance is a viable but clumsy compromise for hierarchists.

The insurance solution even listens and responds to this rationality. Fatalists neglect any strategic approach to achieve LATER. Nonetheless, due to the obligation to contract insurances, they will finally be insured – even if they have to pay very high premiums. But in the end, the damage that they probably suffer does not burden the community.

Finally, the described scheme provides responsiveness and accessibility for all rationalities. Responding to the clumsy floodplains leads to a clumsy land policy.

Everlasting Clumsiness

Finally, a land policy for extreme floods must respond to the clumsy floodplains. This response is necessary, because the clumsy floodplain produces a technological lock-in into a levee-based flood protection. Levee-based flood protection, however, is not able to cope with extreme floods, moreover, it entrenches the lock-in to higher and higher levees until a technology like LATER cannot be applied.

As mentioned earlier, the landscape of clumsiness consists rather of columns than of a slope. However, for simplification, user-friendly Cultural Theory allows to depict the landscape as a slope, as long as the contemplator is aware of the fact that the slope is not mathematically linear increasing continuously. The slope allows the following thoughts.

We found the answer to the clumsy floodplains: an obligatory insurance against natural hazards. But clumsiness is not a static concept. Rationalities are in a permanent dynamic imbalance. They respond to new situations. In the course of time, when LATER is applied in almost each large catchment area, the land policy mobilises LATER very niftily. The competition between the rationalities becomes less competitive; the land policy becomes more and more institutionalised. Rationalities lose interest in having access to the policy process. The land policy becomes less clumsy.

This happened to the Tullaian River Rhine reconstructions. When Johann Gottfried von Tulla reconstructed the Rhine, it was probably a clumsy solution. Today, however, the clumsy solution became a closed hegemony – water managers just build levees today without considering other rationalities than prescribed in the administrative procedures. So, the clumsy floodplains emerged. Individualists started to build houses in floodplains, egalitarians defend these houses, etc.

The landscape of clumsiness is not as static as it seems. Once a clumsy solution is achieved, it does not necessarily stay as clumsy as it is. This means, a polyrational solution might become less polyrational in the course of time. Policy processes strive at monorationality. Probably, fewer rationalities raise their voice due to time, and policy responds to fewer rationalities. If the policy is represented as a ball in this landscape (see Figure 3.11), due to time, it rolls down the slope (indeed, it bounces from one column to another, from a high level of clumsiness to a lower level). For different policy problems, the slope might look different. Since

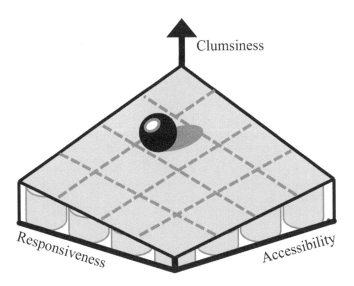

Figure 3.11   Landscape of clumsiness and Sisyphean policymaking

the slope is a very abstract illustration of policy processes, it will be difficult to draw an exact and empirically proved slope. Nonetheless, the inclination will be similar in different policy processes.

Clumsy policymaking must therefore strive at keeping the ball uphill. This is like a Sisyphean task:

> Aye, and I saw Sisyphus in violent torment, seeking to raise a monstrous stone with both his hands. Verily he would brace himself with hands and feet, and thrust the stone toward the crest of a hill, but as often as he was about to heave it over the top, the weight would turn it back, and then down again to the plain would come rolling the ruthless stone. But he would strain again and thrust it back, and the sweat flowed down from his limbs, and dust rose up from his head (Homer, *Odyssey*, book 11, lines 593–600).

Policymaking, indeed society, has permanently to aim at granting access to all rationalities and try to respond to all rationalities. Then, a responsive land policy emerges. It responds to clumsiness. An obligatory insurance against natural hazards is a response to the clumsy floodplains. It is a socially constructed answer to extreme floods. Once responsive land policy is implemented, each rationality within to the social construction of floodplains should contribute to keep the ball uphill. This means each rationality should maintain clumsiness by insisting on its ideals. By a conflict of ideals, a clumsy – but viable – land policy for extreme floods can be achieved.

Bibliography

Aachener Zeitung 2008. Schutz gegen Hochwasser soll in städtischer Hand bleiben. [Online, 25 September]. Available at: www.az-web.de/lokales/heinsberg-detail-az/666015 [accessed 25 September 2008].

Alterman, Rachelle 2007. More than land assembly: Land readjustment for the supply of urban public services, in *Analyzing Land Readjustment – Economics, Law, and Collective Action*, edited by Yu-Hung Hong and Barrie Needham. Cambridge, Massachusetts: Lincoln Institute of Land Policy, 57–88.

Arellano, A.L., Vetere, A. de Roo and J.-P. Nordvik 2007. Reflections on the challenges of EU policy making with view to flood risk management, in *Flood Risk Management in Europe – Innovation in Policy and Practice*, edited by Selina Begum, Marcel J.F. Stive and Jim W. Hall. Dordrecht: Springer, 433–68.

Arthur, W. Brian 1989. Competing technologies, increasing returns, and lock-in by historical events, *The Economic Journal*, 99, 116–31.

Assmann, André 2001. Dezentraler, integrierter Hochwasserschutz – vom Konzept zur Planung, in *Hochwasserschutz Heute – Nachhaltiges Wassermanagement*, edited by Stefanie Heiden, Rainer Erb and Friedhelm Sieker. Berlin: Erich Schmidt Verlag, 197–236.

Bahlburg, Cord Heinrich 2005. Hochwasser und andere Katastrophen – Was haben wir gelernt?, in *Risiken in Umwelt und Technik – Vorsorge durch Raumplanung: Forschungs- und Sitzungsberichte Band 223*, edited by Helmut Karl, Jürgen Pohl and Horst Zimmermann. Hannover: Akademie für Raumforschung und Landesplanung, 3–14.

Battis, Ulrich, Michael Krautzberger and Rolf-Peter Löhr 2007. *Baugesetzbuch, Kommentar*. 10th edition. München: C.H. Beck.

Baumol, William J. and Wallace E. Oates 1988. *The Theory of Environmental Policy*. 2nd edition. Cambridge: Cambridge University Press.

Beder, Sharon 2006. *Environmental Principles and Policies: An Interdisciplinary Approach*. London: University of New South Wales Press Ltd.

Begum, Selina, Marcel J.F. Stive and Jim W. Hall 2007. *Flood Risk Management in Europe – Innovation in Policy and Practice*. Dordrecht: Springer.

Berendes, Konrad 2008. Kommentar zum Gesetz zur Ordnung des Wasserhaushalts, in *Das Deutsche Bundesrecht*. München: C.H. Beck.

Berg, Marco, Georg Erdmann, Markus Hofmann, Michael Jaggy, Martin Scheringer and Hansjörg Seiler 1994. *Was ist ein Schaden? Zur Normativen Dimension des Schadensbegriffs in der Risikowissenschaft*. Zurich: vdf editors.

Bezirksregierung Cologne 2003. *Planfeststellungsbeschluss Köln-Porz-Langel. Planfeststellungsbeschluss vom 20.11.2003 für den Abschnitt 12 Lülsdorf/ Langel*. Köln: Bezirksregierung Köln.

Billé, Raphael 2008. Integrated coastal zone management: Four entrenched illusions, *Surveys and Perspectives Integrating Environment and Society*, 1, 75–86.

BMVBS (Federal Ministry for Transport, Building, and Urban Affairs) 2006. *Hochwasserschutzfibel – Bauliche Schutz- und Vorsorgemaßnahmen in hochwassergefährdeten Gebieten*. Bonn: BMVBS.

Boettcher, Roland 1997. Wird die Umweltplanung dem Problem Hochwasser gerecht?, in *Hochwasser – Natur im Überfluss?*, edited by Ralf Immendorf. Heidelberg: C.F. Müller Verlag, 115–44.

Böhm, Hans R. and Michael Deneke 1992. *Wasser – Eine Einführung in die Umweltwissenschaften*. Darmstadt: Wissenschaftliche Buchgesellschaft.

Bollens, Scott, Edward Kaiser and Raymond Burby 1988. Evaluating the effects of local floodplain management policies on property owner behaviour, *Environmental Management*, 12 (3), 311–25.

Botzen, W.J.W., J.C.J.H. Aerts and J.C.J.M. van den Bergh 2009. Willingness of homeowners to mitigate climate risk through insurance, in *Ecological Economics* [Online]. Available at: http://www.sciencedirect.com/ [accessed 18 March 2009].

Breuer, Rüdiger 2006. Die neuen wasserrechtlichen Instrumente des Hochwasserschutzgesetzes vom 3.5.2005, *Natur und Recht*, 28 (10), 614–23.

Bromley, Daniel W. 1991. *Environment and Economy – Property Rights and Public Policy*. Oxford: Basil Blackwell Ltd.

Brown, James D. and Sarah L. Damery 2002. Managing flood risk in the UK: Towards an integration of social and technical perspectives, *Transactions of the Institute of British Geographers – New Series*, 27 (4), 412–26.

Browne, Mark J. and Robert E. Hoyt 2000. The demand for flood insurance: Empirical evidence, *Journal of Risk and Uncertainty*, 20 (3), 291–306.

Buitelaar, Edwin 2003. Neither market nor government – Comparing performance of user right regimes, *Town Planning Review*, 74 (3), 315–30.

City of Cologne 1996. *Hochwasserschutzkonzept Köln*. Cologne: Dezernat Bauen und Verkehr.

Coase, Ronald 1960. The problem of social cost, *Journal of Law and Economics*, 3, 1–44.

Cooley, Heather 2006. Floods and draughts, in *The World's Water*, edited by Peter H. Gleick. Washington: Islandpress, 91–116.

Cooter, Robert and Thomas Ulen 2004. *Law and Economics*. International edition. 4th edition. Boston, San Francisco, New York and others: Pearson Addison Wesley.

Corell, Cathrin 1996. Schaffung und Bewahrung von Retentionsräumen zum Zwecke des Hochwasserschutzes, *Umwelt und Planungsrecht*, 7, 246–53.

Cormann, Petra 2008. Kommentar zum WHG, in *Umweltrecht – Beck'scher Online Kommentar*, edited by Ludger Giesberts and Michael Reinhardt. München: C.H. Beck.

Czychowski, Manfred 1998. *Wasserhaushaltsgesetz – Kommentar*. München: Verlag C.H. Beck.

Davy, Benjamin 1997. *Essential Injustice – When Legal Institutions Cannot Resolve Environmental and Land Use Disputes*. Wien: Springer.

Davy, Benjamin 1999. Boden und planung – Zwischen privateigentum und staatsintervention, in *Dortmunder Beiträge zur Raumplanung 89: Was ist Raumplanung*, edited by Klaus M. Schmals. Dortmund: Institut für Raumplanung, 101–19.

Davy, Benjamin 2004. *Die Neunte Stadt – Wilde Grenzen und Städteregion Ruhr2030*. Dortmund: Müller+Bussmann.

Davy, Benjamin 2005. Bodenpolitik, in *Handwörterbuch der Raumordnung*, 4th edition, edited by Ernst Hasso Ritter. Hannover: Akademie für Raumforschung und Landesplanung, 117–30.

Davy, Benjamin 2006. *Innovationspotentiale für Flächenentwicklung in schrumpfenden Städten: Flächenmanagement am Beispiel Magdeburgs* [Online]. Available at: www.iba-stadtumbau.de/index.php?Innovationspotent iale-fur-Flachenentwicklung-in-schrumpfenden-Stadten-1 [accessed 21 April 2009].

Davy, Benjamin 2007. Mandatory happiness? Land readjustment and property in Germany, in *Analyzing Land Readjustment – Economics, Law, and Collective Action*, edited by Yu-Hung Hong and Barrie Needham. Cambridge, Massachusetts: Lincoln Institute of Land Policy, 37–55.

Davy, Benjamin 2008. Plan it without a condom!, *Planning Theory*, 7 (3), 301–17.

Die Rheinpfalz 29 June 2007. Der Hochwasserschutz ist ins Wasser gefallen, *Die Rheinpfalz* [regional newspaper], 148.

Dieterich, Hartmut 2000. *Baulandumlegung*. 4th edition. München: Verlag C.H. Beck.

DKKV (Deutsches Komitee für Katastrophenvorsorge e.V.) 2003. *Hochwasservorsorge in Deutschland – Lernen aus der Katastrophe 2002 im Elbgebiet. Kurzfassung der Studie*. Bonn: Schriftenreihe des DKKV.

Doe, Robert 2006. *Extreme Floods – A History in a Changing Climate*. Phoenix Mill, Gloucestershire: Sutton Publishing.

Douglas, Mary 2005. *Grid and Group, New Developments* [Online: Paper on the workshop on complexity and Cultural Theory in honour of Michael Thompson, London School of Economics and Political Science]. Available at: www.psych. lse.ac.uk/complexity/Workshops/MaryDouglas.pdf [accessed 9 March 2007].

Dransfeld, Egbert and Frank Osterhage 2003. *Einwohnerveränderungen und Gemeindefinanzen*. Dortmund: Institut für Landes- und Stadtentwicklungsforschung des Landes NRW.

Düsterdiek, Bernd 2001. Vom regen in die traufe? Was kommunen zum schutz vor hochwasser leisten können, *KA Wasserwirtschaft, Abwasser, Abfall*, 48 (9), 1201–2.

Ellis, Richard and Michael Thompson 1997. *Culture Matters – Essays in Honor of Aaron Wildavsky*. Oxford: Westview Press.

Encyclopedia Britannica 2009. Fundamental attribution error [Online]. Available at: ww.britannica.com/EBchecked/topic/222156/fundamental-attribution-error [accessed 22 February 2009].

Engel, H., P. Kraheé, U. Nicodemus, P. Heininger, J. Pelzer, M. Disse and K. Wilke 2002. *Das Augusthochwasser 2002 im Elbegebiet*. Koblenz: Bundesanstalt für Gewässerkunde.

Epping, Volker and Christian Hillgruber 2008. *GG – Beck'scher Online Kommentar*. München: C.H. Beck.

Ermer, Klaus 2001. Grenzüberschreitende zusammenarbeit beim hochwasserschutz, *Schriftenreihe des Deutschen Rates für Landespflege*, 72, 83–7.

ESDP 1999. *European Spatial Development Perspective – Towards Balanced and Sustainable Development of the Territory of the European Union*. Brussels: European Commission.

European Union 2007. Directive 2007/60/EC of the European Parliament and of the Council of 23 October 2007 on the assessment and management of flood risks, *Official Journal of the European Union*, L 288, 27–34.

Fleischhauer, Mark 2004. Klimawandel, Klimagefahren und Raumplanung; Dortmunder Vertrieb für Bau- und Planungsliteratur, Dortmund.

Gerdes, Frank 2008. Wiederherstellung von Retentionsflächen – Fallanalyse der Retentionsflächen Köln-Langel und Köln-Worringen. Dortmund, unpublished diploma thesis.

German Bundestag 1995. *Drucksache 13/1207: Entwurf eines Gesetzes zur Änderung des Wasserhaushaltsgesetzes (WHG)*. German Bundestag.

German Bundestag 1996. *Drucksache 13/4788: Beschlußempfehlung und Bericht des Ausschusses für Umwelt, Naturschutz und Reaktorsicherheit Entwurf eines Gesetzes zur Änderung des Wasserhaushaltsgesetzes (WHG)*. German Bundestag.

German Bundestag 1998. *Drucksache 13/11092: Antwort der Bundesregierung auf die Große Anfrage der Abgeordneten Gila Altmann (Aurich), Ulrike Höfken, Steffi Lemke, weiterer Abgeordneter und der Fraktion BÜNDNIS 90/DIE GRÜNEN – Drucksache 13/9466 – Flußausbaumaßnahmen und Hochwassergefahr in der Bundesrepublik Deutschland*. German Bundestag.

German Bundestag 2002. *Drucksache 14/9894: Gesetzentwurf der Fraktionen SPD und BÜNDNIS 90 / DIE GRÜNEN. Entwurf eines Gesetzes zur Änderung steuerrechtlicher Vorschriften und zur Errichtung eines Fonds 'Aufbauhilfe' (Flutopfersolidaritätsgesetz)*. German Bundestag.

German Bundestag 2005. *Drucksache 15/3168: Gesetzentwurf der Bundesregierung: Entwurf eines Gesetzes zur Verbesserung des vorbeugenden Hochwasserschutzes*. German Bundestag.

German Bundestag 2009. *Drucksache 16/12786: Entwurf eines Gesetzes zur Neuregelung des Wasserrechts*. German Bundestag.

German Government 2005. *Bericht der Bundesregierung über die nach der Flusskonferenz vom 15. September 2002 eingeleiteten Maßnahmen zur Verbesserung des vorbeugenden Hochwasserschutzes*. Berlin: Bürgerservice des Bundesministeriums für Verkehr, Bau- und Wohnungswesen.

Gieseke, Paul, Werner Wiedemann and Manfred Czychowski 1992. *Wasserhaushaltsgesetz – Kommentar*. München: Verlag C.H. Beck.

Greiving, Stefan 2002. *Räumliche Planung und Risiko*. München: Gerling Akademie Verlag.

Greiving, Stefan 2003. Ansatzpunkte für ein Risikomanagement in der Raumplanung, in *Raumorientiertes Risikomanagement in Technik und Umwelt – Katastrophenvorsorge durch Raumplanung. Forschungs- und Sitzungsberichte Band 220*, edited by Helmut Karl and Jürgen Pohl. Hannover: Akademie für Raumforschung und Landesplanung, 114–31.

Greiving, Stefan 2006. Dealing with natural hazards in Germany's planning practice, in *Natural Hazards and Spatial Planning in Europe*, edited by Mark Fleischhauer, Stefan Greiving and Sylvia Wanczura. Dortmund: Dortmunder Vertrieb für Bau- und Planungsliteratur, 55–76.

Grünewald, Uwe 2005. Vom Hochwasser-'Schutzversprechen' zum 'Hochwasser-Risikomanagement', in *Hochwassermanagement*, edited by Robert Jüpner. Magdeburg: Shaker Verlag, 5–22.

Hartmann, Thomas 2005. Hochwasserschutz durch räumliche Planung – Handlungsempfehlungen für Abstimmungsprozesse zwischen den Akteuren in Sachsen-Anhalt. Diploma thesis at TU Dortmund.

Hartkopf, Günter and Eberhard Bohne 1983. *Umweltpolitik, Band 1 – Grundlagen, Analysen, Perspektiven*. Opladen: Westdeutscher Verlag.

Haupter, Birgit and Peter Heiland 2002. Vorsorgender Hochwasserschutz: Verantwortung der Raumordnung?, *RaumPlanung*, 104, 233–5.

Haupter, Birgit, Peter Heiland and J. Neumüller 2007. Interregional and transnational co-operation in river basins – chances to improve flood risk management, in *Flood Risk Management in Europe – Innovation in Policy and Practice*, edited by Selina Begum, Marcel J.F. Stive and Jim W. Hall. Dordrecht: Springer, 505–22.

HdWW, Bd. 2 1988. *Handwörterbuch der Wirtschaftswissenschaft. Vol. 2*. Stuttgart and New York: UTB.

Heemeyer, Carsten 2007. *Auswirkungen des Hochwasserschutzgesetzes auf Raumordnungs- und Bauleitpläne*. Berlin: Lexxion.

Heiland, Peter 2002. *Vorsorgender Hochwasserschutz durch Raumordnung, interregionale Kooperation und ökonomischen Lastenausgleich*. Darmstadt: Verein zur Förderung des Institutes WAR.

Helbig, Bernd 2003. *Die Hochwasserkatastrophe im August 2002 in Dessau*. Dessau: Rupa-Druck.

Hill, Peter J. and Roger E. Meiners 1998. *Who Owns the Environment?* Lanham, Maryland: Rowman & Littlefield Publishers.

Homer, *Odyssey* [Online]. Available at: www.perseus.tufts.edu [accessed 1 May 2009].

Huber, Michael 2004. Insurability and regulatory reform: Is the English flood insurance right able to adapt to climate change?, *The Geneva Papers on Risk and Insurance*, 29 (2), 169–82.

IKSE (International Commission for the Protection of the Elbe River) 2003. *Aktionsplan Hochwasserschutz Elbe*. Magdeburg: IKSE.

IKSR (International Commission for the Protection of the Rhine) 1998. *Aktionsplan Hochwasser*. Koblenz: IKSR.

IKSR (International Commission for the Protection of the Rhine) 1999. *Wirkungsabschätzung von Wasserrückhalt im Einzugsgebiet des Rheins*. Koblenz: IKSR.

IKSR (International Commission for the Protection of the Rhine) 2001. *Rhein 2020 – Programm zur nachhaltigen Entwicklung des Rheins*. Koblenz: IKSR.

IKSR (International Commission for the Protection of the Rhine) 2002. *Hochwasservorsorge – Maßnahmen und ihre Wirksamkeit*. Koblenz: IKSR.

IKSR (International Commission for the Protection of the Rhine) 2005. *Flood Action Plans 1995–2005 – Action Targets, Implementation and Results*. Koblenz: IKSR.

IPCC (Intergovernmental Panel on Climate Change) 2001. *Climate Change 2001: Synthesis Report, Contribution of the Working Group I, II, and III to the Third Assessment Report of the IPCC*. Cambridge: Cambridge University Press.

IPCC (Intergovernmental Panel on Climate Change) 2007. *Fourth Assessment Report of the Intergovernmental Panel on Climate Change – Summary for Policymakers* [Online]. Available at: www.ipcc.ch [accessed 20 September 2007].

Janssen, Gerold 2005. Hochwasserschutz, in *Handwörterbuch der Raumordnung*, 4th edition, edited by Ernst Hasso Ritter. Hannover: Akademie für Raumforschung und Landesplanung, 451–6.

Johnson, Clare L. and Sally J. Priest 2008. Flood risk management in England: A changing landscape of risk responsibility, *Water Resources Development, Theme Issue: The Public-Private Divide in Flood Management*, 24 (4), 513–26.

Karl, Helmut and Jürgen Pohl 2003. *Raumorientiertes Risikomanagement in Technik und Umwelt – Katastrophenvorsorge durch Raumplanung: Forschungs- und Sitzungsberichte Band 220*. Hannover: Akademie für Raumforschung und Landesplanung.

Kotulla, Michael 2006. Das gesetz zur verbesserung des vorbeugenden hochwasserschutzes, *Neue Zeitschrift für Verwaltungsrecht*, 25 (2), 129–35.

Kotulla, Michael 2007. *Wasserhaushaltsgesetz zwischen Hochwasserschutz, Strategischer Umweltprüfung und Föderalismusreform*. Stuttgart: Kohlhammer Verlag.

Krautzberger, Michael 2007. *Baugesetzbuch – Beck'scher Online Kommentar*. Version 86. München: C.H. Beck.

Krendelsberger, Wolfram 1996. *Handelbare Belastungsrechte in der Umweltpolitik*. Wien: Manz'sche Verlags- und Universitätsbuchhandlung.

Kron, Wolfgang 2003a. Hochwasserrisiko und Überschwemmungsvorsorge in Flussauen, in *Raumorientiertes Risikomanagement in Technik und Umwelt – Katastrophenvorsorge durch Raumplanung. Forschungs- und Sitzungsberichte Band 220* edited by Helmut Karl and Jürgen Pohl. Hannover: Akademie für Raumforschung und Landesplanung, 79–101.

Kron, Wolfgang 2003b. Hochwasser und uberschwemmungen: Bekämpfen oder akzeptieren?, *Schadensspiegel*, 46 (3), 26–35.

Kron, Wolfgang 2004. Hochwasser und uberschwemmungen: Bekämpfen oder akzeptieren?, *DWA-Rundbrief Landesverband Bayern*, 2, 9–14.

Land Trust Alliance 2007. *Website of the Land Trust Alliance* [Online]. Available at: www.lta.org [accessed 12 February 2007].

Lane, Jon 2006. Foreword, in *The World's Water*, edited by Peter H. Gleick. Washington, DC: Islandpress.

Lange, Jan 2005. Property-Rights-Verträge im Rahmen des Risikomanagements von Elementarschäden, *German Risk and Insurance Review*, 1 (5), 153–72.

LAWA (Interstate Working Group on Water) 1995. *Leitlinien für einen zukunftsweisenden Hochwasserschutz*. Berlin: Geschäftsstelle der Länderarbeitsgemeinschaft Wasser.

LAWA (Interstate Working Group on Water) 2004. *Instrumente und Handlungsempfehlungen zur Umsetzung der Leitlinien für einen zukunftsweisenden Hochwasserschutz*. Berlin: Geschäftsstelle der Länderarbeitsgemeinschaft Wasser.

Lecher, Kurt, Gerd Lange and Herbert Grubinger 2001. Gewässerregelung, in *Taschenbuch der Wasserwirtschaft*, edited by Kurt Lecher, Hans-Peter Lühr and Ulrich C. Zanke. Berlin: Parey Verlag.

Loucks, Daniel P., Jery R. Stedinger, Darryl W. Davis and Eugene Z. Stakhiv 2008. Private and public responses to flood risks, *Water Resources Development, Theme Issue: The Public–Private Divide in Flood Management*, 24 (4), 541–54.

Luhmann, Hans-Jochen 2005. Soziale dämme vor folgen des klimawandels – Auf Treibsand gebaut, *Zeitschrift für Rechtspolitik*, 38 (1), 22–4.

Mamadouh, Virginie 1999. Grid-group cultural theory: An introduction, *GeoJournal*, 47 (3), 395–409.

Ministerie van Verkeer en Waterstaat 2009. Staatssecretaris start debat over noodoverloopgebieden [Online]. Available at: www.verkeerenwaterstaat.nl/ [accessed 18 February 2009].

Ministry for Agriculture and the Environment of Saxony-Anhalt 2003. *Hochwasserschutzkonzeption des Landes Saxony-Anhalt bis 2010*. Magdeburg.

Ministry for the Environment and Transport of Baden-Württemberg, Ministry of Interior of Baden Württemberg and Ministry of Economics of Baden-

Württemberg 2003. *Hochwassergefahr und Strategien zur Schadensminderung in Baden-Württemberg.* Stuttgart.

MKRO 1998. *Raumordnung und vorbeugender Hochwasserschutz.* Decision of the ministers responsible for regional planning. 4 June 1998.

Moss, Timothy and Jochen Monstadt 2008. *Restoring the Floodplains in Europe – Policy Contexts and Project Experiences.* London: IWA Publishing.

Munich Re Group 2003a. Katastrophenportrait: Die sommerüberschwemmungen in Eurpa – ein Jahrtausendhochwasser?, *Munich Re Group: Topics – Jahresrückblick Naturkatastrophen 2002,* 10, 16–25.

Munich Re Group 2003b. Volkswirtschaftliche Konsequenzen der August-Überschwemmungen in Deutschland – eine Bestandsaufnahme, *Munich Re Group: Topics – Jahresrückblick Naturkatastrophen 2002,* 10, 26–31.

Munich Re Group 2005. *Wetterkatastrophen und Klimawandel – Sind wir noch zu retten?* Edition Wissen. Munich: Munich Re Group.

Needham, Barrie 2006. *Planning, Law and Economics – The Rules We Make for Using Land.* London and New York: Routledge.

Needham, Barrie 2007a. *Dutch Land Use Planning – Planning and Managing Land Use in the Netherlands, the Principles and the Practice.* Den Haag: Sdu uitgevers.

Needham, Barrie 2007b. The search for greater efficiency: Land readjustment in the Netherlands, in *Analyzing Land Readjustment – Economics, Law, and Collective Action,* edited by Yu-Hung Hong and Barrie Needham. Cambridge, Massachusetts: Lincoln Institute of Land Policy, 115–34.

Needham, Barrie 2007c. Final comment: Land use planning and the law, *Planning Theory,* 6 (2), 183–9.

Ney, Steve 2007. *Messy issues, Policy Conflict and the Differentiated Polity: Analysing Contemporary Policy Responses to Complex, Uncertain and Transversal Policy Problems* [Online]. Available at: https://bora.uib.no/bitstream/1956/1512/1/Thesis-Ney.pdf [accessed 2 February 2009].

NJW 1957. Sozialgebundenheit des Grundeigentums: Grünfläche. BGH judgement from 20 December 1956, III ZR 82/55 (Hamm), *Neue Juristische Wochenschrift,* 14, 538–9.

Patt, Heinz 2001. *Hochwasser-Handbuch – Auswirkungen und Schutz.* Berlin: Springer.

Paul, Matthias and Jutta Pfeil 2006. Hochwasserschutz in der bauleitplanung (unter besonderer Berücksichtigung des Hochwasserschutzgesetzes 2005), *Neue Zeitschrift für Verwaltungsrecht,* 25 (5), 505–12.

Petrow, Theresia, Annegret H. Thieke, Heidi Kreibich, Cord Heinrich Bahlburg and Bruno Merz 2006. Improvements on flood alleviation in Germany: Lessons learned from the Elbe flood in August 2002, *Environmental Management,* 38 (5), 717–32.

Potter, Karen 2008. *Planning Space for Water.* Proceedings of the 10th BHS National Hydrology Symposium: Sustainable Hydrology for the 21st Century, University of Exeter, 15–17 September 2008, 341–5.

Random House Webster's Dictionary 1993. *School and Office Dictionary*. New York: Random House.

Rawls, John 2005. *A Theory of Justice*. (Reprint, originally published in 1971). Cambridge, Massachusetts and London: The Belknap Press of Harvard University Press.

Reinhardt, Michael 2003. Retentionsflächen und eigentum: Zur wasserrechtlichen Planfeststellung von Deichverlegungen, *Zeitschrift für Wasserrecht*, 42 (4), 193–212.

Reinhardt, Michael 2004. Hochwasserschutz zwischen Enteigungsentschädigung und Amtshaftung, *Natur und Recht*, 26 (7), 420–29.

Reinhardt, Michael 2008. Der neue eurpäische hochwasserschutz, *Natur und Recht*, 30 (7), 468–73.

Rolfsen, M. 2009. Das neue wasserhaushaltsgesetz, *NuR*, 31, 765–71.

Roth, Dik and Jeroen Warner 2007. Flood risk uncertainty and changing river protection policy in the Netherlands: The case of 'calamity polders', *Tijdschrift voor Economische en Sociale Geografie*, 98 (4), 519–25.

Rötzsch, Matthias 2005. Entwicklung von Ansätzen zur Minderwertberechung für Objekte in Hochwassereinflussgebieten, *Der Sachverständige*, 32 (12), 376–9.

Scheele, Ulrich 2006. Privatisierung, liberalisierung und deregulierung in netzgebunden infrastruktursektoren, in *Wandel der Stromversorgung und Räumliche Politik*. Forschungs- und Sitzungsberichte der ARL No. 227. Edited by Dieter Gust. Hannover: Akademie für Raumforschung und Landesplanung.

Scherer, Donald (ed.) 1990. *Upstream/Downstream – Issues in Environmental Ethics*. Philadelphia, Pennsylvania: Temple University Press.

Schneider, Sandra 2005. *Rechtliche Instrumente des Hochwasserschutzes in Deutschland*. Berlin: Erich Schmidt.

Schönberger, Rainer 2005. Hochwasserschutz und versicherung, in *Rechtliche Aspekte des vorbeugenden Hochwasserschutzes*, edited by Wolfgang Köck. Baden-Baden: Nomos, 193–9.

Schwarz, Michel and Thompson, Michael 1990. *Divided We Stand – Redefining Politics, Technology and Social Choice*. New York, London, Toronto, Sydney and Tokyo: Harvester Wheatsheaf.

Schwarze, Reimund and Gerd G. Wagner 2004. In the aftermath of Dresden: New directions in German floods insurance, *The Geneva Papers on Risk and Insurance*, 29 (2), 154–68.

Segeren, Arno, Femke Verwest, Barrie Needham and Edwin Buitelaar 2007. Property rights and private initiatives: An introduction, *Town Planning Review – Special Issue: Property Rights and Private Initiatives*, 78 (1), 9–22.

Seher, Walter 2004. Hochwasserschutz – Handlungsoptionen der raumplanung zwischen koexistenz und kooperation, *Wasserwirtschaft*, 94 (3), 8–12.

Sieder, Frank, Herbert Zeitler, Heinz Dahme and Günther-Michael Knopp 2007. *Wasserhaushaltsgesetz, Abwasserabgabengesetz – Kommentar*. 33rd edition. München: C.H. Beck.

Spieth, Friedrich 2008. Kommentar zum WHG, in *Umweltrecht – Beck'scher Online Kommentar*, edited by Ludger Giesberts and Michael Reinhardt. München: C.H. Beck.

Staatliches Umweltamt Krefeld 2002. *Jeder cm zählt – Hochwasser(schutz) am Niederrhein* (DVD and booklet). Krefeld: Staatliches Umweltamt Krefeld.

Steenhoff, Holger 2003. Rechtliche instrumente des hochwasserschutzes, *Umwelt- und Planungsrecht*, 23 (2), 50–6.

Strobl, Theodor and Frank Zunic 2006. *Wasserbau – Aktuelle Grundlagen*. Berlin and Heidelberg: Springer.

Stüer, Bernhard 2004. Hochwasserschutz im spannungsverhältnis zum übrigen fachplanungsrecht, raumordnungsrecht und zur bauleitplanung, *Natur und Recht*, 26 (7), 415–20.

Stüer, Bernhard 2005. *Handbuch des Bau- und Fachplanungsrechts*. 3rd edition. München: C.H. Beck.

Swiss Re 1998. *Floods – An Insurable Risk?* Zürich: Swiss Reinsurance Company.

Swiss Re 2002. *Floods are Insurable!* Zürich: Swiss Reinsurance Company.

Thompson, Michael 2008a. *Democratic Governance, Technological Change and Globalisation*. Workshop report of the science in society symposium by the Economic and Social Research Council (ESRC), Lisbon, February 2007.

Thompson, Michael 2008b. *Organising and Disorganising – A Dynamic and Non-linear Theory of Institutional Emergence and its Implications*. AxmInsert: Triarchy Press.

Thompson, Michael, Richard Ellis and Aaron Wildavsky 1990. *Cultural Theory*. Oxford: Westview Press.

Thompson, Michael, Gunnar Grenstedt and Per Selle 1999. *Cultural Theory as Political Science*. London: Routledge.

UBA (Umweltbundesamt) 1998. *Ursachen der hochwasserentstehung und ihre anthropogene beeinflussung*. UBA Texte 18/98. Berlin: UBA.

UBA (Umweltbundesamt) 1999. *Anforderungen des vorsorgenden Hochwasserschutzes an Raumordnung, Landes-/Regionalplanung, Stadtplanung und die Umweltfachplanungen – Empfehlungen für die Weiterentwicklung*. UBA Texte 45/99. Berlin: UBA.

UBA (Umweltbundesamt) 2003. *Sicherung und Wiederherstellen von Hochwasserrückhalteflächen*. UBA Texte 34/03. Berlin: UBA.

USACE (US Army Corps of Engineers), FEMA (Federal Emergency Management Agency), ASFPM (Association of State Floodplain Managers) and NAFSMA (National Association of Flood and Stormwater Management Agencies) 2009. National flood risk management program [Online]. Available at: www.iwr. usace.army.mil/nfrmp/index.cfm [accessed 5 March 2009].

Van Rij, Evelien 2008. *Improving Institutions for Green Landscapes in Metropolitan Areas*. Amsterdam: Delft University Press.

Verweij, Marco 2000. *Transboundary Environmental Problems and Cultural Theory – The Protection of the Rhine and the Great Lakes*. New York: Palgrave.

Verweij, Marco, Michael Thompson and Christoph Engel 2006b. Clumsy solutions: How to do policy and research in a complex world, in *Clumsy Solutions for a Complex World – Governance, Politics and Plural Perceptions* edited by Marco Verweij and Michael Thompson. Hampshire and New York: Palgrave Macmillan, 241–9.

Verweij, Marco, Mary Douglas, Richard Ellis, Christoph Engel, Frank Hendriks, Susanne Lohmann, Steven Ney, Steve Rayner and Michael Thompson 2006a. The case for clumsiness, in *Clumsy Solutions for a Complex World – Governance, Politics and Plural Perceptions*, edited by Marco Verweij and Michael Thompson. Hampshire and New York: Palgrave Macmillan, 1–27.

Vischer, Daniel and Andreas Huber 1993. *Wasserbau*. 5th edition. Berlin: Springer.

Voigt, Manfred 2005. Hochwassermanagement und räumliche Planung, in *Hochwassermanagement*, edited by Robert Jüpner. Magdeburg: Shaker Verlag, 97–117.

Welzer, Harald 2009. *Unbesorgt am Abgrund – Wie Menschen sich die Wirklichkeit ihren Illusionen anpassen*. Radio report at NDR Kultur, 22 March 2009, 19:00–19:15.

Wesselink, Anna 2007. Flood safety in the Netherlands: the Dutch response to hurricane Katrina, *Technology in Society*, 29 (2), 239–47.

Zehetmair, Swen, Jürgen Pohl, Katharina Ehrler, Britta Wöllecke, Uwe Grünewald, Sabine Mertsch, Reinhard Vogt and Yvonne Wieczorrek 2008. Hochwasservorsorge und Hochwasserbewältigung in unterschiedlicher regionaler und akteursbezogener Ausprägung, *Hydrologie und Wasserbewirtschaftung*, 52 (4), 203–11.

Zwanikken, Timm 2001. Ruimte als voorraad? Ruimte als variëteit! – De consequenties van discoursen rondom 'ruimte als voorraad' voor het rijks ruimtelijke beleid [Online]. Available at: http://webdoc.ubn.kun.nl/mono/z/ zwanikken_t/ruimalvo.pdf [accessed 22 April 2009].

Interviews

The interviews were conducted by Thomas Hartmann as semi-structured interviews. The landowners have been randomly selected in flood-prone areas. Interview protocols are on file with the author.

Landowners in Biederitz, near Magdeburg. Three interviews. Biederitz, 30 May 2008.

Landowners in Bitterfeld. Three interviews. Bitterfeld, 31 May 2008.

Landowners in Cologne–Rodenkirchen. Six interviews. Cologne–Rodenkirchen, 13 May 2008.

Landowners in Dessau-Waldersee. Four interviews. Dessau-Waldersee, 31 May 2008.

Landowners in Torgau. Four interviews. Torgau, 1 June 2008.

Mayor of Bitterfeld. Interview with Dr. Werner Rauball. Bitterfeld, 1 June 2008.

Mayor of Gübs (a small municipality near Magdeburg). Interview with Karl-Heinz Latz. Gübs, 30 May 2008.

Officer at the Ministry of Building and Transport Saxony-Anhalt. Interview with Frank Thäger. Magdeburg, 20 June 2008.

Officers of the municipality of Dessau-Rosslau. Interview with Ingolf Schmidt (town planning) and Mr Mardicke (environmental department). Dessau--Rosslau, 20 June 2008.

Regional planner of the Region Magdeburg. Interview with Eckhard Groß. Magdeburg, 30 May 2008.

Thompson, Michael. Interview. Via e-mail, 3 April 2009.

Town planner of Magdeburg. Interview with Hans-Joachim Olbricht. Magdeburg, 30 May 2008.

Water manager at state agency for flood protection (Landesbetrieb für Hochwasserschutz), of Saxony-Anhalt. Interview with Gerd Dörre. Lutherstadt-Wittenberge, 30 November 2004.

Water managers at state agency for flood protection (Landesbetrieb für Hochwasserschutz) of Saxony-Anhalt. Interview with Gerd Dörre and Peter Noack, Lutherstadt-Wittenberge, 30 May 2008.

Index